Multiple Stressors in the Environment

Multiple Stressors in the Environment

Assessing Ecological Effects in Complex Systems

Edited by

Lawrence A. Kapustka
LK Consultancy
Diamond Valley, AB, Canada

Wayne G. Landis
Western Washington University
Bellingham, US

David J. Ostrach
Ostrach Consulting
Woodland, US

Cameron Irvine
Robertson-Bryan, Inc.
Gold River, US

WILEY

Published by John Wiley & Sons, Inc., Hoboken, New Jersey.

Library of Congress Cataloging-in-Publication Data applied for:

Hardback ISBN: 9781119886044

Cover image(s): Visual Haiku, Sound of Water 1 by Susan L. Kristoferson
Cover design: Wiley

Set in 9.5/12.5pt STIXTwoText by Straive, Chennai, India

Contents

List of Figures *xi*
List of Tables *xvii*
List of Contributors *xix*
About the Editors *xxi*
Prologue *xxv*
Abbreviations *xxxi*

1 **Impetus for This Book** *1*
David J. Ostrach, Lawrence A. Kapustka, and Cameron A. Irvine
References *4*

2 **Foundations of Environmental Science** *5*
Lawrence A. Kapustka
References *11*

3 **Exposures and Effects** *13*
Cameron A. Irvine, Mary A. Ottinger, and Lawrence A. Kapustka
Introduction *13*
Analytical Tools *14*
Toxicity Testing *17*
Predictive Toxicology *21*
Increasingly Sensitive Endpoints and Expanding Knowledge *23*
The Exposome and One Health *25*
References *28*

4 Sequence Effects *35*
Lawrence A. Kapustka
References *41*

5 Addressing the Challenges of Pharmaceutical Pollution *47*
Sam A. Welch
Introduction *47*
Drivers of Pharmaceutical Risk *48*
  Demographic and Market Change *50*
  Health and Management Change *51*
  Climate Change *52*
Modeling the Risks of Pollutant Mixtures *52*
Modeling the Risks of Multiple Stressors *54*
Bayesian Networks for Environmental Risk Assessment *54*
BNs for ERA: A Norwegian Example *55*
Future Outlooks *59*
Conclusions *61*
Acknowledgments *61*
References *61*

6 The Role of Contaminants, in the Context of Multiple Stressors, in the Collapse of a Striped Bass Population *71*
David J. Ostrach and Cameron A. Irvine
Introduction *71*
Approach *71*
  Contaminant Exposures *72*
Biochemical Activation *73*
  EROD *74*
  Immunointegrity *75*
  Acetylcholinesterase Inhibition *75*
  Metallothionein *77*
  Vitellogenin *77*
  Biochemical Effects Synthesis *79*
Egg-Ovaries *79*
Developmental and Histopathology Effects *80*
Otoliths – Habitat Use *82*
Conclusion *83*
Acknowledgments *83*
References *83*

7 **Management of Environmental Systems Using Cumulative Risk Assessment – The Importance of Multiple Stressor Cumulative Ecological Risk Assessment for Long-Term Management of Environmental Systems Demonstrated Using the Upper San Francisco Estuary Case Study** *91*
Wayne G. Landis
Introduction *91*
Basics of Using the RRM for Multiple Stressor Risk Assessment *91*
The Use of BNs for Calculation *92*
The Upper San Francisco Estuary Example – The Study Site *94*
The Development of the Conceptual Model *96*
Pesticide Data *98*
Water Quality Data *98*
Chinook, Delta Smelt Trawl Data *98*
Additional Chemical Screening and Geographic Distribution *100*
Building the Predicted Fish Mortality Pathway *100*
Options for the Conditional Probability Tables of the Predicted Fish Mortality Node *104*
Calculation of Risk to the Individual Species *106*
Discussion: Estimating Risk with Multiple Stressors/Cumulative Risk Assessment *108*
References *111*

8 **Integrating Emerging Science in Advancing Risk Assessment for Wildlife** *115*
Mark S. Johnson and Thomas G. Bean
Introduction *115*
Considering Data Generated from New and Emerging Tools in Problem Formulation *117*
Improved Methods for Exposure Estimation *117*
Aspirational Ways to Improve Accuracy of Exposure Models *123*
What Has Been Incorporated Recently into Exposure Models? *126*
Future Options for Improving Exposure Assessment Models *127*
Novel Tools for Estimating Hazard (Toxicity) *131*
In vivo *Alternative* Tools (Adverse Outcome Pathways, Cell Lines, IVIVE) *131*
Methodological Improvements in Current Guidelines and Laboratory Model Development *134*

Development of Toxicological Benchmarks Using Evidence Integration Techniques *136*
Research and Data Needs *137*
References *138*

9 Reflections from an Environmental Toxicologist Working in a Large Company *147*
Ralph G. Stahl Jr (Retired)
Introduction *147*
Remediation and Restoration of Contaminated Land in the United States *148*
Remediation and Restoration of Contaminated Land Outside the United States *148*
What Is a "Site" *149*
DuPont Remediation Program *150*
Risk-Based Decision-Making *151*
Sites/Examples *151*
The United States – Superfund Site, Newport, Delaware *151*
The United States – RCRA-Corrective Action Site, Waynesboro, Virginia *153*
The United Kingdom – Contaminated Lands Program Site, Northern Ireland *155*
Discussion *157*
Acknowledgments *158*
Conflict of Interest Statement *159*
Disclaimer *159*
Data Availability *159*
References *159*

10 Big Data – Artificial Intelligence and Machine Learning – Examples of the Application of Approaches Derived from Artificial Intelligence/Machine Learning to Analyze Large Datasets for Ecological Risk Assessment *163*
Wayne G. Landis
Introduction *163*
Section 1: Microcosm/Mesocosm Summary, the Effects of Jet Fuels, and Community Conditioning *165*
Section 2: Ecological Risk Assessment/Relative Risk Model and the Port Valdez Risk Assessment *170*
Section 3: Transition to BNs *175*

Section 4: Case Summary: Predicting Risk to Estuary Water Quality and Patterns of Benthic Environmental DNA in Queensland, Australia *176*
Section 5: Adaptive Management and Next Steps *179*
Cautions/Caveats *180*
References *181*

11 New Tools and Approaches in Ecological Risk Assessment *185*
P. Bruce Duncan and Jeanmarie M. Zodrow
Introduction *185*
Despite the Goal to Advance Ecological Risk Assessment, Progress Can Be Very Slow *185*
Lessons from Development and Use of NAMs to Reduce Animal Testing *188*
Framework for Tool Development *192*
Actions to Advance the Use of New Tools and Approaches *195*
Q1. Why Are Uses in Decisions Not Easily Found? *195*
Q2. How Do You Unlock Obstacles to Advancement? *196*
Q2-Suggested Actions *197*
Q3. What Creates Compelling Arguments for Decision-Makers? *200*
Q3-Suggested Actions *201*
Q4. What Support for Tools Is Needed for Success? *203*
Q4-Suggested Actions *203*
Q5. Where Do We See Successes? *205*
Q5-Suggested Actions *205*
Summary and Conclusions *209*
Checklist of Actions *210*
References *213*
Appendix: Results of Author Survey *217*

12 Going Forward *221*
Lawrence A. Kapustka
Introduction *221*
References *224*

Index *227*

List of Figures

Figure 2.1 Mismatch between stated protection goals and the methods used to assess them *10*

Figure 3.1 The essential ingredients that affect individual human health (Exposome) are interactive and involve the environment (One Health) *27*

Figure 4.1 Conceptual model depicting the complexity of the social-ecological system influencing Sage Grouse *40*

Figure 5.1 A conceptual, nonexhaustive diagram of the first-order influence of health issues on pharmaceutical consumption (a), routes of pharmaceutical emission to the environment (b), and environmental impacts (c); and potential future drivers of change: demographic/market, health/management, and climate change. Original diagram, based on *49*

Figure 5.2 (a) Bayesian networks are directed acyclic graphs (DAGs), i.e., the relationships between nodes have direction, and nodes and edges cannot form a loop. (b) Variables are represented as nodes with different possible states (or a continuous distribution of state) that have different probabilities. In the example, the state of Node 2 depends on the state of Node 1. The specific probabilities of these are depicted in a Conditional Probability Table (CPT). When a Bayesian network is compiled, the probability of each state of every node is calculated iteratively using these CPTs and the states of prior nodes. (c) Object-oriented Bayesian networks allow repeated motifs of nodes (represented by rounded rectangles) to be deployed in subnetworks *55*

Figure 5.3 Condensed diagram of an object-oriented Bayesian network used for the prediction of pharmaceutical environmental risk. The main network (a) consists of a number of scenario nodes, six API (active pharmaceutical ingredient) exposure and risk calculation subnetworks corresponding to low, medium, and high sales weight APIs, and 10 risk nodes (orange) that calculate both sums of Risk Quotients (RQs) and probability of exceeding various RQ thresholds, respectively, by category and by number of APIs. Rounded rectangles indicate instances of subnetworks, containing API information and a pipeline for forecasting sales weights and predicting environmental concentrations and subsequent risk. Whole-line ovals represent ordinary nodes. Dotted ovals with gray insets represent input nodes, which receive inputs in subnetwork instances from regular nodes. Whole-line ovals with gray insets represent output nodes, which provide outputs from subnetworks to the main network *57*

Figure 5.4 Predicted distributions of Risk Quotients (RQs) for six individual pharmaceuticals and all pharmaceuticals summed across three counties and four combinations of year and population growth scenario *58*

Figure 5.5 Comparison of Risk Quotient (RQ) distributions for six individual pharmaceuticals and all pharmaceuticals summed across three counties and three combinations of wastewater treatment scenario in 2050 *59*

Figure 5.6 A possible future layout of an OOBN for predicting pharmaceutical risk in Norway *60*

Figure 7.1 Conceptual models. (a) The original relative risk model for ecological risk assessment. (b) The basic format populated with site-specific factors for the Upper San Francisco Estuary study area. The conceptual model is the basis for the derivation of the Bayesian network *96*

Figure 7.2 Chemical distribution and concentrations by risk region. The sample sites and the maximum concentrations for chlorpyrifos (a), bifenthrin (b), and mercury (c). Note the distribution of the sampling sites, the variability in maximum chemical concentration, and the variability among the risk regions *101*

Figure 7.3 The Predicted Fish Mortality pathway segment of the Bayesian network *102*

Figure 7.4 Calculation of Predicted Fish Mortality for the Sacramento River Risk Region and the wet season. The calculation is that 52.3% of the time the toxicity would equal to or exceed the EC20. The output is from the additive CPT *103*

Figure 7.5 Sensitivity analysis for the prediction of the fish toxicity output using mutual information. The yellow node is in season. The blue nodes are the chemical concentrations, and the green represents the three pathways. Note that it is the mixtures that are driving the risk *104*

Figure 7.6 Comparison of outputs between the CPT options for the Confluence Risk Region. The Additive CPT has the highest total toxicity, the Average CPT the lowest, and the Expert Judgement the broader distribution *105*

Figure 7.7 The difference in sensitivity between the Additive (a) and the Expert Judgement derived (b) models by risk region. Both models have similar sensitivities to nodes except that the Expert Judgement model had greater mutual information values in each case *107*

Figure 7.8 The difference in sensitivity between the Additive (a) and the Expert Judgement (b) derived models by wet or dry season. Both models have similar sensitivities to nodes except that the Expert Judgement model had greater mutual information values in each case *108*

Figure 7.9 Risk calculation for each species. In (a) each of the risk regions is selected giving an overlook of the risk. In (b) the Suisan Bay risk region is selected. Note the similarity in the two distributions for each species *109*

Figure 8.1 Conceptual diagram of factors potentially influencing exposure and effects in birds *119*

Figure 8.2 Simplified schematic of current approaches to pesticide exposure assessment for birds and mammals *120*

Figure 8.3 Simplified tiered approach for acute and chronic exposure to pesticides for birds and mammals *121*

Figure 8.4 Schematic building upon ideas presented in Morrissey et al. (2023) to incorporate chemical, environmental, organismal, and ecological complexity into exposure assessment based on European Food Safety Authority (2023) exposure assessment tiers. Note Table 2 is taken from Morrissey et al. (2023) *124*

Figure 8.5 Individual mean ratio of exposure of lead in soil to Eastern Bluebirds compared to the EcoSoil screening level lead toxicity reference value (1.63 mg/kg-d) for birds for a lead exposure within a small arms range. Model assumed 15 foraging event/day and a population size of 16 birds for a nesting season of 150 days *130*

Figure 8.6 Schematic illustrating how in vivo alternative tools (in vitro, in silico, and in ovo) could complement existing hazard assessments for birds *133*

Figure 10.1 Standard aquatic microcosm (SAM) and mixed flask culture (MFC) multispecies toxicity test methods. (a) is a depiction of the initial stages of the experiment after the containers have been inoculated with the species listed in the protocol and randomized in the experimental room. (b) depicts a time in the experiment with the algae blooming due to the lack of Daphnia because of the toxicity. (c) depicts the arrangement of the individual MFC containers *166*

Figure 10.2 Multivariate analysis of the impact of Jet-A in the standardized aquatic microcosm (SAM) test system. (a) shows the Cosine distance from the control group to each of the treatments for each sampling day. Note that large differences are apparent early in the SAM. During the middle part of the 63-day experiment, the distances between the replicates of Treatment 1, the control group, is as large as the distances to the treatment groups. However, later in the experiment, the distances from the dosed microcosms to the control again increase. Significance levels of the three multivariate statistical tests for each sampling day are presented in (b) *168*

Figure 10.3 Trajectories of different treatments of JP-8 and heat in a Taub microcosm. The pentagon marks day 63 of a typical study. (a) is a space-time worm projection using *Anabena* and small amphipods as the variables. (b) depicts the trajectories for *Paramecium bursaria* and *Ankistrodesmus*. Note how the two trajectories provide information about each system at different times of the experiment. At certain periods of the experiments, the distance along these axes are small between treatment groups but can then change as the treatments age. Key: Reference are nondosed (Treatment 1), heat only (Treatment 2), JP-8 (Treatment 3), and JP-8/heat (Treatment 4) *169*

Figure 10.4 The basic format of the relative risk model (RRM). This format is still the basis of the current Bayesian network-based RRMs *172*

Figure 10.5 Bayesian network (BN) model example for the Logan estuary. A model was built for each of the study areas. This is the BN for the Lower and Middle part of the Logan with the Middle Logan selected for the risk calculation *178*

Figure 11.1 Key elements to obtain broader acceptance of the new approaches and application to higher tier regulatory decisions (EPA CompTox blueprint) *190*

Figure 11.2 Example of systematic process with specified outcome *193*

Figure 11.3 Bridging from case studies to decision-making in the use of tools/approaches by decision-makers *194*

List of Tables

Table 1.1 SETAC sessions on multiple stressors 2012–2025 *3*

Table 2.1 Limitations of ecological risk assessments *8*

Table 7.1 Chemicals considered for the Predicted Fish Mortality pathway *99*

Table 10.1 Examples of ecological risk assessments using the relative risk mode (RRM), probabilistic relative risk model (PRRM) using Monte Carlo, and the Bayesian network–relative risk model (BN-RRM) as noted in the text *171*

Table 10.2 Probability of meeting management objectives to two endpoints by each of the risk regions *179*

Table 11.1 Key roles for government, industry, the regulatory sector, academia, and journals in development and acceptance of new approach methodologies *191*

List of Contributors

Thomas G. Bean
FMC Corporation
Avian Toxicology
Newark
DE
USA

P. Bruce Duncan
Bruce Duncan Environmental Consulting
Boulder
CO
USA

Cameron A. Irvine
Associate Environmental Scientist at Robertson-Bryan, Inc.
Environment Department
Rancho Cordova
CA
USA

Mark S. Johnson
Defense Centers for Public Health – Aberdeen, Toxicology Directorate
Aberdeen Proving Ground
MD
USA

Lawrence A. Kapustka
LK Consultancy
Diamond Valley
AB
Canada

Wayne G. Landis
College of the Environment, Institute of Environmental Toxicology and Chemistry
Western Washington University
Bellingham
WA 98229
USA

David J. Ostrach
Chief Scientist at Ostrach Consulting
Woodland
CA
USA

Mary A. Ottinger
Professor Emeritus and Research Professor, Department of Biology and Biochemistry
University of Houston
Houston
TX
USA

Ralph G. Stahl Jr
DuPont Company
Corporate Environmental
Affairs and Environmental Toxicology
Wilmington
DE
USA

Samuel A. Welch
Norwegian Institute for Water
Research (NIVA)
Ecotoxicology and Risk Assessment
Oslo
Norway

Jeanmarie M. Zodrow
Geosyntec Consultants
Risk Assessment and Applied
Toxicology
Greenwood Village
CO
USA

About the Editors

Lawrence (Larry) A. Kapustka is an Emeritus Senior Ecologist and a SETAC Fellow with experience as a professor, government research scientist, and an environmental consultant. Although mostly retired, he work as an independent consultant in the areas of ecotoxicology, ecological risk assessment, ecosystem services, and sustainability. With 40+ years of experience in basic and applied ecological aspects of environmental assessment and management, his work has extended across a wide range of topics involving natural sciences and environmental policy.

After receiving a doctorate in 1975 from the University of Oklahoma, Larry worked as a university professor at the University of Wisconsin–Superior and Miami University, Oxford, Ohio. Later, he worked for nearly three years at the US Environmental Protection Agency (USEPA) Laboratory in Corvallis, Oregon, as lead of the Plant Toxicology and Hazardous Waste teams. Since 1991, he has been a consultant in a company he formed (***e***cological ***p***lanning and ***t***oxicology, Inc.) and then with Golder Associates and SLR Consulting in Calgary. On a limited basis, he continues to consult and offer training in ecotoxicology, ecological risk assessment, ecosystem services, water quality issues, and sustainability and also provides expert witness testimony in contested cases.

Larry is a founding member of the Foothills Energy Co-op and the Diamond Valley Sustainable Living Centre for which he serves as Chair. Beyond the technical fields in which he has published more than 100 peer-reviewed papers, chapters, and books, he authored *Gifts for Santa: Collected Dreams and Confessions* and co-produced *Vistas of the West*, a book of poetry and art. He is the co-editor of *Wisdom River: Meditations on Fly Fishing and Life Midstream* that includes an essay, *Line to Table, Capturing the Total Fishing Experience,* and has contributed an essay *Challenging Times* in *Reimagining Fire: The Future of Energy*.

Wayne G. Landis served as Director of the Institute of Environmental Toxicology and Chemistry from September 1989 until December 2021. He is now a Research Professor within the institute that is part of what is now the College of the Environment. He graduated with a BA in Biology from Wake Forest University (1974) and his MA (1978) and Ph.D. (1979) in Zoology from Indiana University. His first research position was as a Research Biologist, Toxicologist for the Chemical Research and Engineering Center of the US Army at Aberdeen Proving Ground, Maryland. In this position, he was awarded two US patents for the degradation of chemical warfare agents. At WWU, his research and teaching focused on toxicology, population biology, and risk assessment. He has been strong supporter of graduate program in Huxley and of undergraduate research. Since arriving at WWU, his group has published over 200 papers, government technical reports, textbooks, and research topic books, and, as of fall 2025, he has been the advisor for 43 students that have received an MS in Environmental Sciences and has served on the MS or PhD committees of 22 others at WWU and across the world. He taught three Honors courses on the history of the formulation of evolution by natural selection and the scientific and cultural impacts of synthetic biology brought about by CRISPR technology. He is now Research Professor Emeritus and conducting research on the risk assessment of microplastics, contaminants, and synthetic biology and on the use of risk assessment to evaluate resource damage and remediation.

David J. Ostrach received his bachelor's degree in Marine Biology and Neuroethology from University of California (UC) San Diego and his PhD in Comparative Pathology from the School of Veterinary Medicine at UC Davis.

David's research career began at Scripps Institution of Oceanography investigating microbial degradation of crude oil spills prior to the construction of the Alaska pipeline. This was followed by working on food chain research investigating population growth, feeding, development, and rejection behavior in the copepod *Calanus pacificus*. He took a position as Associate Research Scientist and Manager of the Aquatic Toxicology Laboratory at Science Applications Inc., JRB Associates, Environmental Chemistry and Geochemistry Division in La Jolla, CA. In that position, he was responsible for the design and development of an aquatic toxicology facility including the drafting of original laboratory system plan for static and flow-through testing as well as marine and freshwater organism culture.

David started and led an environmental consulting firm, Bioaquatic Consultants, that specialized in the development of protocols for performance of compliance monitoring, testing for aquatic toxicology, and terrestrial impact studies. He then entered the aquaculture business by starting the Lazy O Fish Ranch in Descanso, CA, a 28-acre aquaculture facility involved in the production and sale of various species of fish, algae, and plankton as well as various organisms for use in aquatic toxicology testing/bioassays.

David then took a position at UC Davis, where he worked on various environmental and cancer research studies from 1987 to 2009. He designed and supervised the construction of the Aquatic Toxicology Research Laboratory Facility building at UC Davis. As its first director, he developed and implemented inland surface water toxicity testing along with state-of-the-art environmental research programs. The main focus of his laboratory at UC Davis was to conduct interdisciplinary research on striped bass and factors associated with its long-term population decline. Recently, his laboratory has been involved in the Interagency Ecological Program (IEP) efforts to determine factors responsible for the Pelagic Organism Decline in the San Francisco Estuary.

David's laboratory at UC Davis, the Pathobiology, Conservation and Population Biology Laboratory, conducted interdisciplinary research on sublethal effects of contaminants on striped bass *Morone saxatilis* in the San Francisco Estuary system. His overall program goal was to better understand the long-term effects of contaminant exposure, climate change and multiple stressors on fish populations, ecosystem health, and human health.

David is currently Chief Scientist at Ostrach Consulting providing interdisciplinary and basic research services focusing on aquatic ecosystems, pathobiology, aquatic toxicology, risk assessment, remediation/restoration as well as fisheries, aquaculture, aquaponics, and sustainable technology. His research goals are to better understand the long-term effects of contaminant exposure and multiple stressors on fish and wildlife populations, ecosystem health, and to correlate findings with human health and risk assessment.

Cameron A. Irvine is an Associate Environmental Scientist and Ecotoxicologist with Robertson-Bryan, Inc. in Rancho Cordova, California, USA. He earned his Bachelor of Science at the University of Saskatchewan in 1994 and a Master of Science focused on ecotoxicology from Wright State University in 2003. Between receiving these degrees, he was an environmental consultant in Canada, conducting environmental assessments at gold and uranium mining sites in northern Saskatchewan and leading fisheries studies and stream restoration projects focused on salmonids in central British Columbia.

A career as an environmental consultant has allowed Cameron to participate in a broad range of studies throughout North America and in Europe where he supported a range of clients from federal agencies to first nations, industries, and municipal utilities. Cameron has led investigations focused on the effects of contamination on aquatic resources, including ecological risk assessments at Superfund sites, pilot remediation studies, and toxicity-reduction evaluations that have identified novel toxicants at municipal waste treatment facilities. He has shared these experiences and lessons learned as a member of the Society of Environmental Toxicology and Chemists (SETAC) for over 25 years and has had the pleasure of cochairing a session on the effects from multiple stressors with his coeditors of this book at the annual SETAC North American conference for the past 10 years.

Cameron has contributed to the scientific community by chairing several SETAC committees, and he received the Eugene Kenaga Membership Award from SETAC in 2020. He also chairs the ASTM International E50.47 subcommittee on Biological Effects and Environmental Fate, which develops and maintains international toxicity testing standards. He has coauthored several peer-reviewed scientific publications related to ecotoxicology and ecological risk assessment, led numerous scientific investigations, authored many reports, and coauthored a chapter on *Radionuclides in Biota* in the book *Environmental Contaminants in Biota* (2nd Edition, Beyer and Meador Eds., 2011, CRC Press).

Prologue

This book provides an in-depth exploration of interactions of multiple stressors within complex social-ecological systems. The introductory chapters chart the foundation of environmental assessment and management and set the stage for consideration of the effects of various combinations of biological, chemical, and physical stressors.

A brief history of the modern environmental movement provides context for our modern challenges. This begins with an overview of the state of the environment during the turbulent 1960s in the United States that was dominated by protests of the US–Viet Nam war, a vibrant civil rights movement, a push for an equal rights amendment that would give women their rightful status in society in the United States, and continuation of the Cold War between the United States and the Soviet Union. Elsewhere there was turmoil in Central and South America, Africa, and the Middle East. During this time, the Cuyahoga River notoriously was so polluted with organic effluents that it caught on fire. Rachael Carson published *Silent Spring*, and the First Earth Day was launched at a UNESCO Conference in San Francisco (1969) and promoted by Wisconsin Senator Gaylord Nelson in 1970.

In response to these events, the US Congress passed the National Environmental Protection Act, the Clean Water Act, the Clean Air Act, and other seminal environmental laws in short order with bipartisan votes. In all, the aspirational features of the laws led to new regulatory frameworks and technological breakthroughs in science.

Key advances occurred in both the sciences and in regulations from that time to now. These include the development of bioassays as an integrating tool to evaluate effects of chemicals and chemical mixtures on select organisms. Parallel tracks led to sophisticated analytical methods in environmental chemistry to quantify substances in water, sediments, soil, air, and biological tissues. These advances occurred before electronic calculators, let alone personal computers, were available. In retrospect, it is remarkable how much was achieved, even though we now recognize how woefully inadequate computational capabilities were in those early days of the modern environmental movement.

Bioassays were designed to measure the effects of chemicals in environmental media. Today's students might be surprised that in the early days analytical chemistry methods were not able to detect, let alone quantify, concentrations of substances in water, sediment, soil, air, or biological tissues. Yet, if known amounts were added to the media, one could demonstrate adverse effects on organisms ranging from suppressed growth and lowered fecundity to death. In many cases, the biological effects could be demonstrated at a nominal concentration that were two or more orders of magnitude lower than could be detected by chemical analytical methods. It was from this state of the science that many in the public sphere came to understand that if a potential toxic material was detectable in the environment, it was already too late; adverse effects were already occurring.

As improvements in analytical tools and methods were developed, we now have the capacity to quantify the concentration of many substances at concentrations well below the physiological response threshold. However, because scientists were so effective in conveying the message that if a substance was detected, effects were already occurring, we find that many in the public have not updated their understanding of toxicological-response relationships and experience great fear of many substances that may be present at *de minimis* concentrations.

On the regulatory side, following the creation of environmental impact assessments, there were new rules and new legal philosophies including the "polluter pays" and "joint and severable" liability in the Superfund legislation, as well as the oil contaminant laws and natural resource damage considerations. The wheels of environmental justice were spinning just as new technological advances were being made. Questions pertaining to relevance, adequacy, equity, and ethics were posed within the sociopolitical milieu. All this, ironically, at a time when trust in the worth of the sciences was being eroded and the use of misinformation and disinformation as a powerful political cudgel was gaining traction.

The tools of the modern environmental movement began to take shape after the passage of the National Environmental Protection Act in the United States in 1970. First out of the chute were guidelines for conducting environmental impact assessments. The initial efforts were focused on documenting what environmental resources were present in the area where some project or environmental policy change was being proposed. Methods borrowed from community ecology, fisheries biology, forestry, wildlife ecology, natural history, and the like were used to generate encyclopedic lists of species known to, or thought to, reside or otherwise use an area. The mandate for the impact assessments was to understand what the project would mean for those ecological resources. The aspiration was that by having a clear indication of how a roadway, dam, or other major development would alter the function of the ecological system, modifications of the project could be made to mitigate the effects. In some cases, the consequences could be determined to be too destructive to win approval. What-if scenarios were intended to describe the current situation (i.e., without the project) and compare it to one or more alternative project designs.[1] The comparisons were intended to inform decision making pertaining to the project: would the benefits to society be greater than the impacts? Despite the laudatory goals, the tools needed to quantify the adverse impacts were woefully inadequate, and as a result most environmental impact assessments were little more than encyclopedic listings of species present, rich in information and mostly silent in terms of ecological or societal consequences. New types of analyses were spawned to estimate environmental and social benefits and costs. The basic concept was that if the benefit-to-cost metrics exceeded 1, the project could be approved. Many projects, for example, new dam constructions, gained positive benefit-to-cost values by adding favorable recreational economic value to push the ratio into positive territory, despite ecologists voicing adverse effects associated with damming riverways.

In the second decade of the modern environmental era, the need for better ways to evaluate adverse effects emerged. The concept of exposure being linked to effects had been described for human health considerations. Building from this, scientists at the Oakridge National Laboratory described how that risk paradigm could be applied to environmental situation for aquatic and terrestrial species. A few years later, in 1989, the US Environmental Protection Agency (USEPA) published the guideline for using risk assessment to evaluate risks at hazardous waste sites. The basic elements of ecological risk assessment were presented in guidance documents by the USEPA. Subsequently, variations of the framework have been adapted in different jurisdictions around the world.

Advances in basic and applied ecology were also underway. Richard Forman, considered the father of landscape ecology, published *Landscape Ecology*[2] with Michael Godron in 1986. This seminal book illustrates how humans interact with nature, the beginnings of framing social-ecological dynamics, and describes how ecological sciences could be considered in terms of spatial patterns. A decade later, James Kay and others championed the concepts of thermodynamics as the driver for self-organization of ecological systems and from this progressed to embracing complexity as a way to understand resiliency in ecological systems. These explorations began to reveal that complex dynamics are such that there is limited capacity to predict future outcomes and that the best we can do is assign crude probabilities to potential alternative outcomes. Additionally, insights of

---

1 In Canada, the approach was a bit different. For the most part, proponents and assessors worked iteratively to mitigate the effects during design phases such that the public were not presented with alternative scenarios.

2 Forman, R.T.T. and Godron, M. (1986). *Landscape Ecology*. New York: Wiley. ISBN 0471870374.

"wicked problems"[3] were being applied to environmental management projects. Together, this awareness led to the recognition that contemporary environmental challenges require what has been defined as postmodern science[4] – a new approach based on the realization that in situations in which the stakes are high, when decisions need to be made quickly, intrinsic uncertainty is high, and there are conflicting values among stakeholders, the traditional science that relies exclusively on cause–effect relationship is insufficient.

The organization of this book is designed to lead the reader through the formative years of the modern environmental movement through seminal developments in both environmental science and policy. This will detail the emergence of single-species bioassays in efforts to characterize the adverse effect of chemicals in the environment, both naturally occurring and those that are released intentionally or accidentally. Emergence of methods to characterize risk plays a major role in how chemicals have been managed. While risk assessment methods were being implemented, critical insights into social-ecological systems have occurred. In many ways, the past two to three decades have been dominated by impressive advances in computational capabilities, molecular biology, and analytical chemistry. We will highlight each of these.

In the chronology of advances, we will focus on limitations that are inherent in assessments that target single chemicals. Obviously, no organism ever encounters only one substance or one stressor, so even though important information can be gained by studying effects of single chemicals under standard settings, much greater sophistication is required to evaluate effects in the complexity of social-ecological systems. We will trace the emergence of efforts to look at mixtures of chemicals as well as combinations of biological, chemical, and physical stressors.

For example, Chapter 6 on the contribution of multiple stressors to the Striped Bass population collapse in the Sacramento-San Joaquin River Delta and San Francisco Bay Estuary explores the application of biomarkers to describe the effects of contaminants in the field. To do this, it compares biomarker responses to the field condition for a number of key species in this critical marine and estuarine environment. In the field of risk assessment, this information would augment the usual type of field information when risk is being estimated for a wide variety of species. Often, we have environmental or tissue concentrations from which effects can be estimated using test data. However, data are lacking on the relationship between many biomarkers and whole-organism-level effects; nevertheless, great strides are being made in the elucidation of adverse outcome pathways (AOPs)–the

---

3 Rittel, H. and Webber, M. (1973). Dilemmas in a general theory of planning. *Policy. Sci.* 4: 155–169.

4 Funtowicz, S.O. and Ravetz, J.R. (1994). Uncertainty, complexity and post-normal science. *Environ. Toxicol. Chem.* 13: 1881–1885.

biochemical pathways linking biomarkers with whole-organism effects. Response data would allow risk assessors to have a confirmation of an effect, reducing the uncertainty of the risk estimate. This information would allow the confirmation or falsification of the risk predictions. One would wish for the day when there are more of this type of data when estimating risks. These data could be applied to the risk assessment of the Upper San Francisco Estuary (Chapter 6) to allow a confirmation of an effect and reduce the uncertainty of the risk estimate.

In an initial risk characterization and interactive model built for the California Delta, also known as the Upper San Francisco Estuary, the endpoints are Chinook Salmon and the Delta Smelt (see Chapter 7). Because of the extensive dataset supplied by the California Environmental Data Exchange Network (CEDEN), it was possible to conduct and extensive investigation of the frequency distribution for a number of contaminants: mercury, bifenthrin, chlorpyrifos, diazinon, myclobutanil, triadimefon, and malathion. A Bayesian network was built to describe the effects of these materials on the percentage of fish mortality. Mercury, bifenthrin, and malathion were the three contaminants with the most influence on forecasting the fish mortality distribution. The interactions of these mixtures were evaluated using an additive model, using the average of the predicted exposure-responses and finally applying expert judgment. The additive model generated the highest probability of toxicity; the expert judgment model had the broadest distribution, and the average model the lowest probability of fish mortality. The integration of tools from machine learning and the use of Bayesian networks make possible the building of risk assessment models that describe the risk due to multiple contaminants to a variety of endpoints.

A key advancement in the investigation of the effects of multiple stressors has been the application of tools developed for machine learning to the analysis of large datasets and the building of predictive models (Chapter 10). In ecological systems, even artificial ones, there are multiple species and in a toxicant such as a petroleum-based fuel, there are a wide number of different hydrocarbons that degrade over time. In the carefully defined microcosm experiments described in this chapter, the different species exhibit several population dynamics, being effected by the hydrocarbons, the ratios of different nutrients and the species interactions. A key result is the demonstration that individual replicates could be identified as to treatment group using multivariate tools able to use datasets to identify their treatment groups better than chance at the p level of 0.05. This was the case for both of the microcosms used, the 3-L standardized aquatic microcosm (SAM, also Taub microcosm) and the 1-L mixed flask culture (MFC). Long-tern studies could resolve treatment-related grouping even after the degradation of the hydrocarbons. These same tools were connected to the application of Bayesian networks to elucidate effects to several estuaries near Brisbane, Australia. To elucidate the numbers of species in each estuary, eDNA was used. The multivariate techniques allowed the identification of patterns in species composition and water quality parameters specific to the estuary. Bayesian networks were constructed that quantified the interactions and allowed predictive models to be used to estimate impacts.

Throughout the book, we will introduce real-world case studies that illustrate how interactions among multiple stressors shape the outcome of environmental consequences that often lead us to markedly different conclusions than we would consider if only single stressor studies were performed. In these case studies, as well as dedicated chapters, we will explore the ways that new technologies show promise of dramatically improving our collective capabilities to assess situations and offer robust environmental management strategies.

Lawrence A. Kapustka, Wayne G. Landis, David J. Ostrach, Cameron A. Irvine

Abbreviations

4-NP	4-nitrophenol
ADME	absorption, distribution, metabolism and excretion parameters or effect
AChE	acetylcholinesterase
AI	artificial intelligence
AIOSA	avian *in ovo* screening assay
AOP	Adverse Outcomes Pathway
AOP-DB	Adverse Outcome Pathway Database
API	Active Pharmaceutical Ingredient
ARRIVE	Animal Research: Reporting of *in vivo* Experiments
ASSI	Area of Special Scientific Interest
ASTM	American Society for Testing and Materials
AVS	acid volatile sulfides
Bay-Delta	San Francisco Bay and Sacramento-San Joaquin Delta
BLM	biotic ligand model
BMD	benchmark dose
BN	Bayesian network
CA	concentration addition
CERCLA	Comprehensive Environmental Response, Compensation, and Liability Act
CRG	Corporate Remediation Group
CEA	cumulative effects assessment
CYP1A	cytochrome P4501A
DDT	dichlorodiphenyltrichloroethane
DEBtox	dynamic energy budget
DNA	deoxyribonucleic acid
DOC	dissolved organic carbon
E2	estradiol
EC10	10 percent effect concentration

EC25	25 percent effect concentration
EEC	estimated environmental concentration
EDC	endocrine disrupting chemicals
EE2	ethinyl estradiol
EFSA	European Food Safety Authority
EqP	equilibrium partitioning
ERA	ecological risk assessment
EROD	7-ethoxyresorufin-O-deethylase
ESB	equilibrium partitioning-based sediment benchmark
EU	European Union
FIFRA	Federal Insecticide, Fungicide, and Rodenticide Act
GC	gas chromatography
GEO	Gene Expression Omnibus
GIS	geographic information system
HPLC	high-performance liquid chromatography
HTP	high throughput
IA	independent action
IBAMA	Brazilian Institute of the Environment and of Renewable Natural Resources
HC	immunohistochemical
ISO	International Organization for Standardization
ITRC	Interstate Technology Regulatory Council
LOAEL/LOAEC	lowest observed adverse effect level/concentration
LOEC	lowest observed adverse effect concentration
MAF	mixture assessment factor
mBLM	multimetal biotic ligand model
MEC	measured environmental concentration
MC-ICP/MS	multicollector ICP mass spectrometer
MLR	multiple linear regression
MS	mass spectrometry
MT	metallothionein
NAM	new approach methods/nonanimal method
NEPA	National Environmental Protection Act
NIEA	Northern Ireland Environment Agency
NOAA	National Oceanic and Atmospheric Administration
NOAEL/NOAEC	no observed adverse effect level/concentration
NOEC	no observed adverse effect concentration
NPL	National Priorities List
OECD	Organization for Economic Cooperation and Development
OOBN	object-oriented Bayesian networks

OP	organophosphate
ORD	Office of Research and Development
ORISE	Oak Ridge Institute for Science and Education
PAH	polycyclic aromatic hydrocarbons
PBDE	polybrominated diphenyl ether
Pbets	physiologically based extraction
PCB	polychlorinated biphenyls
PCDD	polychlorinated dibenzo-p-dioxins
PBK	physiologically based kinetic
PFAS	per- and polyfluoroalkyl substances
PFOA	perfluorooctanoic acid
PFOS	perfluorooctanesulfonic acid
PNEC	predicted no-effect concentration
POD	pelagic organism decline
ppb	parts per billion (e.g., μg/L)
ppm	parts per million (e.g., mg/L)
ppt	parts per trillion (e.g., ng/L)
qPCR	quantitative polymerase chain reaction
QSAR	quantitative structure activity relationship
RCRA	Resource Conservation and Recovery
RCRA-CA	Resource Conservation and Recovery Act Corrective Action
RNS	reactive nitrogen species
ROD	Record of Decision
ROS	reactive oxygen species
RQ	risk quotient
SAC	special area of concern
SEM	simultaneously extracted metals
SETAC	Society of Environmental Toxicology and Chemistry
SeqAPASS	Sequence Alignment to Predict Across Species Susceptibility
Sr	strontium
SRC	Stockholm Resilience Centre
SERDP	Strategic Environmental Research and Development Program
STRAP	Strategic Research Action Plan
TCDD	2,3,7,8-tetrachlorodibenzo-*p*-dioxin
TK/TD	toxicokinetic and toxicodynamic
TRV	toxicity reference value
US	United States
USA	United States of America

USDOI	US Department of the Interior
USEPA	United States Environmental Protection Agency
USFS	US Forest Service
USFWS	US Fish and Wildlife Service
USSEC	US Securities and Exchange Commission
VDEQ	Virginia Department of Environmental Quality
VTG	vitellogenin
WET	whole effluent toxicity
WHO	World Health Organization
WW2	World War II
WWTP	wastewater treatment plant

1

Impetus for This Book

David J. Ostrach[1], Lawrence A. Kapustka[2], and Cameron Irvine[3]

[1] Chief Scientist at Ostrach Consulting, Woodland, CA, USA
[2] LK Consultancy, Diamond Valley, AB, Canada
[3] Associate Environmental Scientist at Robertson-Bryan, Inc., Environment Department, Rancho Cordova, CA, USA

From the beginning of the modern environmental movement, scientists pondered how to best evaluate the environmental effects of chemicals. Reductive approaches were preferred to isolate the putative effects of a given substance on exposed organisms. At the time that bioassays were being developed, it was common to be able to demonstrate an adverse effect on organisms at nominal concentrations, achieved through spikes of known chemical concentrations followed by dilution, that were well below the detection limits of analytical chemistry. It was important to identify which chemicals were responsible for observed fish kills, bird deaths, or dead vegetation. However, ecologists knew that organisms were exposed to multiple chemicals. Additionally, they were aware that organisms were responsive to non-chemical stressors. Plant scientists and veterinarians were aware of the interactions of chemicals and that some, such as copper and molybdenum and selenium and sulfate, had different biological effects than if either were presented independently; ratios of the chemicals were more predictive due to homeostatic dynamics that afforded organisms the ability to modulate physiological activity.

In 1996, the US Environmental Protection Agency recommended moving away from single-stressor analysis. Marshalling data that could be useful for environmental management required more than single-stressor considerations. Even so, the mandate of the Agency was to manage chemicals and, as such, biological and physical stressors were given limited attention. If considered at all, biological and physical stressors were usually addressed only in a section describing uncertainties. Picking up from the recommendation, SETAC organized a Pellston workshop in 1997, where an invited group of global leaders in science gathered to solve this critical environmental challenge, that resulted in the publication of *Multiple Stressors in Ecological Risk and Impact Assessment* (Foran and Ferenc 1999). The book was expanded in 2000, adding *Approaches to Risk Estimation* to the title. Soon after, SETAC organized another Pellston Workshop and published *Ecological Variability: Separating Natural from Anthropogenic Causes of Ecosystem Impairment* (Baird and Burton 2001).

In the latter workshop, the organizers sought to emphasize complex mixtures. Tasked with providing opening remarks, Kapustka sought to distinguish multiple stressors from complex stressors. Put simply, multiple stressors should be considered anytime more than one stressor, whether biological, chemical, or physical, interacted with a receptor. Complex stressors, on the other hand, would be stressors that behave differently in different situations. For example, ammonium interacting with pH or copper interacting with biotic ligands would qualify as complex stressors. Of course, in a field setting, one is likely to deal with both complex and multiple stressors at any given time.

Beginning in 2012, Dr. David Ostrach proposed a session for the Annual Meeting of SETAC North America. The session was approved and in the Long Beach, California meeting that year, the series was launched with the focus on "Assessing Contaminant Effects in Multi-stress Ecosystems: Lessons from and for the San Francisco Bay-Delta Estuary." Each subsequent year, slight revisions to the call for submissions were made, but with the focus on multiple stressors. The call for abstracts for the 2025 SETAC meeting described the problems to be addressed.

"Aquatic ecosystems are impaired throughout the world, and many are in a state of crisis due to multiple stressors where chemical mixtures and interactions with environmental factors impact organism and contribute to population declines. Anthropogenic stressors such as chemical discharges from industry, agriculture, and treated urban wastewater can interact or be compounded by site-specific stressors (e.g., habitat degradation and invasive species) and other general stressors (e.g., temperature and salinity associated with climate change). This is especially true for estuarine ecosystems where tributaries and major river systems converge with a high degree of human activity and where interrelated factors vary spatially and over time. Understanding the role and relative contribution of each stressor contributing to adverse effects is needed to inform management actions but requires a good conceptual understanding of the ecosystem, chemical fate and transport, a significant wide-ranging environmental dataset, a breadth of analytical techniques, and a multidisciplinary approach that invokes an integrated systems perspective. Limited data often poses a challenge to identifying clear stressor–response relationships and robust statistical hypothesis testing. This emphasizes the need for high-quality and extensive data, new data analysis methods, and integrated assessments of responses to multiple stressors. This session will focus on integrated assessments of multiple stressors and analyses of complex environmental data. Presentations will highlight lessons learned in the following areas: (1) research advances in assessing toxic effects in multistress environments, (2) integrated assessment of environments with multiple stressors, and (3) implications for ecosystem management (e.g., case studies)."

The first year of the sessions (2012) had a modest response, yielding only eight submissions. The audience was also relatively small with between 20 and 30 attendees. Subsequent years drew greater interest, reaching a high of 35 accepted presentations and posters in 2019 before the pandemic. Over 13 years, a median of

Table 1.1 SETAC sessions on multiple stressors 2012–2025.

Year	Title	Chairs
2012	Assessing contaminant effects in multistress ecosystems – lessons from and for the San Francisco Bay-Delta estuary	David Ostrach and Stephanie Fong
2013	Assessing contaminant effects in multistress ecosystems	David Ostrach and Larry Kapustka
2014		
2015		Cameron Irvine joined in as a co-chair
2016		David Ostrach, Larry Kapustka, and Cameron Irvine (except in 2018 when Cam was co-chairing the SETAC North America annual meeting program committee)
2017	Assessing contaminant effects in ecosystems with multiple stressors	
2018	Assessing contaminant effects in ecosystems with multiple stressors with a focus on the California Bay-Delta	
2019	Assessing contaminant effects in ecosystems with multiple stressors	
2020	Assessing contaminant effects in ecosystems with multiple stressors	
2021		
2022		
2023		
2024		
2025		

24 presentations were given at this session at the annual SETAC North America meeting. Attendance at the platform sessions also grew as the number of presentations rose, often having more than 100 people in the room, which we considered a good level of interest given the size of the conference and the number of concurrent sessions.

During the past decade and a half, substantive advances in technology and comprehension of systems ecology, ecotoxicology, and dynamics of social-ecological systems have occurred. This book, inspired by the SETAC North American series that dealt with multiple stressors, showcases conceptual and technical advances in the field of environmental assessment and management through expanded reviews and case studies (Table 1.1).

References

Baird, D.J. and Burton, G.A. Jr., (ed.) (2001). *Ecological Variability: Separating Natural from Anthropogenic Causes of Ecosystem Impairment*. Pensacola, FL: SETAC Press.

Foran, J.A. and Ferenc, S.A. (ed.) (1999). *Multiple Stressors in Ecological Risk and Impact Assessment: proceedings from the Pellston Workshop on Multiple Stressors in Ecological Risk and Impact Assessment, 13–18 September 1997, Pellston, Michigan*. Pensacola, FL: SETAC Press.

2

Foundations of Environmental Science

Lawrence A. Kapustka

LK Consultancy, Diamond Valley, AB, Canada

The modern environmental movement arguably was launched with the publication of *Silent Spring* by Rachel Carson in 1962. These were turbulent times with the Cold War between the United States and the Union of Soviet Socialist Republics, the expansion of the US war in Viet Nam, and the assassinations of President John F. Kennedy, Dr. Martin Luther King, Jr., Malcom X, Senator Robert F. Kennedy, and others. Protests seeking civil rights of what at the time was referred to as Negroes and a growing women's rights movement, portrayed as the Equal Rights Amendment, were in the news. It was also a time of greater comity among political factions compared to current relationships as they grappled with the challenges of the times.

The first attempt to establish a national environmental policy (the Resources and Conservation Act) failed to pass in 1959, but the debate for environmental regulations had begun. The Clean Air Act was passed and signed into law in 1963. The decade-long movement to articulate a national policy culminated in the introduction of the National Environmental Policy Act (NEPA) by Senator Henry Jackson of Washington on 18 February 1969 and was passed in the Senate on 10 July 1969 by a unanimous vote! The House of Representatives voted on its version of the Act on 23 September 1969, passed by a vote of 372-15. The final votes after going through Conference were taken by the Senate on 20 December and the House on 23 December 1969. President Richard M. Nixon signed the bill into law on 1 January 1970, thereby setting in motion an energized environmental agenda (for more detail, see National Environmental Policy Act – Ballotpedia, accessed 11 February 2023). The NEPA created the Council on Environmental Quality and led to the formation of the US Environmental Protection Agency (USEPA).

The vision articulated in NEPA was to declare a national policy, which would encourage productive and enjoyable harmony between humans and the environment; to promote efforts to prevent or eliminate damage to the environment and biosphere and stimulate the health and welfare of humans; and to enrich the understanding of the ecological systems and natural resources important to the nation. The initial efforts to implement NEPA were grandiose attempts to fashion policies and actions that concurrently addressed social and environmental issues. The broad-based, holistic approaches quickly faded with the realization that neither the social sciences nor the biological sciences were equipped with methods to describe the dynamics of human behavior. Practitioners wanted clear, definitive guidance on how to address contaminated waters, poisoned soils and sediments, and fouled air. In short order, engineers came to dominate environmental management; if for no other reason, they were accustomed to offering focused solutions to specific problems. Although many engineers are adept at troubleshooting systems, there was always a strong push for simple direct fixes. If there was a problem of flooding, construct a dam to regulate the flow of water. If the cost exceeded the direct benefit, then recreational value of the soon-to-be-created water body was added, which tipped the ratio favorably. Toxic spills were handled by digging up the contaminated soil or sediment and hauling it to a repository, even if this meant destruction of valuable wildlife habitat. Linear approaches came to dominate environmental management.

The national environmental policy was stated in Sec. 101 [42 USC § 4331] of NEPA[1]:

> The Congress, recognizing the profound impact of [hu]man's activity on the interrelations of all components of the natural environment, particularly the profound influences of population growth, high-density urbanization, industrial expansion, resource exploitation, and new and expanding technological advances and recognizing further the critical importance of restoring and maintaining environmental quality to the overall welfare and development of [hu]man[s], declares that it is the continuing policy of the Federal Government, in cooperation with State and local governments, and other concerned public and private organizations, to use all practicable means and measures, including financial and technical assistance, in a manner calculated to foster and promote the general welfare, to create and maintain conditions under which [hu]man[s] and nature can exist in productive harmony, and fulfill the social, economic, and other requirements of present and future generations of Americans.

1 The full act can be viewed at NEPA of 1969 (energy.gov) (accessed January 2023).

Acting quickly, the Congress and the Administration passed several companion laws that were intended to reverse the decline in environmental quality resulting from pollution that was so prominent in that era, including the Clean Air Act (1970), the Water Quality Improvement Act (1970), the Water Pollution and Control Act Amendments (1972), the Resource Recovery Act (1970), the Resource Conservation and Recovery Act (1976), the Toxic Substances Control Act (1976), the Occupational Safety and Health Act (1970), the Federal Environmental Pesticide Control Act (1972), the Endangered Species Act (1973), the Safe Drinking Water Act (1974), the Federal Land Policy and Management Act (1976), and the Surface Mining Control and Reclamation Act (1977). Similar lists of legislative acts were passed in countries throughout the world over the next two decades.

Translating policy into action was more difficult than many imagined, in large part, because knowledge of the systems and their functions was woefully incomplete. Linear approaches to environmental management often bear unintended consequences. For example, if the focus is solely to eliminate contaminants, then it is likely that critical habitat for the valued wildlife is likely to incur collateral damage that is detrimental to the stated protection goals. However, it is not surprising that the initial efforts were found to be wanting. Only recently is society in general becoming aware of the importance of adopting a landscape perspective that relies on holistic considerations of alternative management scenarios (See Kapustka and Landis 2010).

The effects of environmental pollutants in the late 1960s were for the most part direct observations of dead or deformed fish and wildlife or of problems with domesticated livestock.[2] Human health issues were also being recognized, especially with aminotriazole-tainted cranberries[3] in November 1959. Analytical chemistry methods in that era were by today's standards exceptionally crude and unable to quantify concentrations of most toxic substances even when additions of the substances to water, sediment, or soil in nominal concentrations could be shown to harm organisms. This generated a meme *If you can detect a substance in the environment, it is too late; it has already caused harm.*

As the modern environmental movement began, scientists struggled for ways to describe the effects of pollutants, how to mitigate the adverse effects of pollutants, and how to hold parties responsible for the problems pollutants created. Given the inadequacies of analytical chemistry at the time, bioassays were developed and promoted to address the problems.

Whereas the bioassays or toxicity tests provided basic data, the tool most used to assess the consequences to social-ecological systems has been ecological risk assessment, first promoted for superfund sites in 1989 (Warren-Hicks et al. 1989)

---

2 This was not a new phenomenon. In the early 1900s, ranchers in Montana sued mining companies for damages to livestock (primarily horses) that were dying due to fluoride poisoning.
3 See The Great Cranberry Scare of 1959 | The New Yorker (accessed 11 February 2023).

and more broadly with the framework of 1992 (USEPA 1992). These procedures sought to incorporate inherent toxicity information from bioassays with exposure under environmental conditions. Implementation of ecological risk assessment, however, was slow. A peer review of 17 risk assessments selected by the USEPA was conducted at a workshop in Seattle in the 1990s (personal experience as one of the panelists). Of the 17, only one assessment was judged to have followed the framework, and the remediation selected was not one that was recommended in the assessment. For many of the assessments, there was insufficient description to know the ecological setting of the site (i.e., urban, rural, forest, grassland, riparian, wetland, agricultural field, industrial site, and so on).

Soon after the framework was published, critiques of the overall process were raised. The critiques ranged from inadequate methods to evaluate wildlife exposures, poor alignment of data with stated protection goals, reliance on test data that had low statistical power, use of supposed control sites to determine the magnitude of effects, poor sampling designs, and much more. In 2006, I was invited to speak on the limitations of ecological risk assessments at a USEPA symposium on Risk Assessment in Washington, D.C. I elaborated the core of my presentation into my 2008 publication on limitations (Kapustka 2008). In this analysis, I identified seven inherent and seven contrived limitations for robust ecological risk assessments (Table 2.1). In the19 years since I formulated the lists, ironically, more has been done to address the inherent limitations than the contrived ones. Progress on dealing with the inherent limitations has been substantial. For example,

- Enhanced computing capacity to deal with big data and to perform complex modeling routines, including exploration of probability of alternative outcomes, provide deeper understanding of stochasticity and causality in various scenarios.
- Refinement and speed of methods to detect and map environmental DNA and functionality through sRNA further advance our collective ability to interpret ecological processes.

Table 2.1 Limitations of ecological risk assessments.

Inherent	Contrived
1. Stochasticity	1. Legal/regulatory
2. Scale [space, time]	2. Policy/precedent
3. Baseline ecological information	3. Ecotheocracy
4. Toxicity profiles	4. Point estimates
5. Exposure conditions	5. Data quality
6. Multiple stressors	6. Perceived value/perceived cost
7. Complex stressors	7. Trustworthiness

- Though used sparingly, landscape perspectives and recognition of the importance of thinking in terms of complex social-ecological systems provide a better foundation for engaging in meaningful dialogue regarding management options to identify and deal with risks.

The contrived limitations, however, have stubbornly continued, even though they could easily be addressed. For example,

- There continues to be too many cases (that is $n > 0$) in which no observable adverse effects concentrations (NOAECs) and lowest observable adverse effects concentrations (LOAECs) are used;
- Risk quotients are still being used despite their inherent limitations (nonscalable) and, frequently, even added to project cumulative risk; and
- Risks posed by individual chemicals are seldom considered in the context of other stresses including habitat quality and land use practices that may be more consequential.

A workshop exploring the state of risk assessments for wildlife was concluded in 2022 (Sample et al. 2022). Rather disconcerting was that many of the criticisms and shortcomings of risk assessment posed in the 1990s remain largely unaddressed. A trend toward commoditization of risk leads away from using advancements in the relevant science. Also, there is wide acceptance of default to standard practices even when they can be recognized as inadequate for the problems being addressed.

Bréchignac et al. (2016) and Bradshaw et al. (2014) described the mismatch between stated protection goals and the methods used to evaluate risks (Figure 2.1). Although their focus was on radiation safety, the basic arguments are applicable across the spectrum of ecological risk assessment.

The language in the US NEPA appeared to foster holistic approaches to environmental management. It prescribed a multidisciplinary focus to evaluate environmental impacts of projects and activities that would significantly affect the human environment. So why have so many environmental management problems emerged? A critical impediment is embedded in the linear approach adopted by most regulatory guidelines and followed diligently by practitioners. The next chapter and accompanying case studies trace the historical challenges faced while trying to examine stressor interactions that add greater ecological relevance to environmental assessments. We examine the decade-plus effort to gain acceptance for research on multiple stressors.

But as we address multiple stressors, we would be wise to take stock of where we are as we have passed six decades since *Silent Spring*. The global human population in 1962 totaled 3.12 billion (Population of WORLD 1962 – PopulationPyramid.net, accessed 29 May 2023). Today, the human population exceeds 8 billion

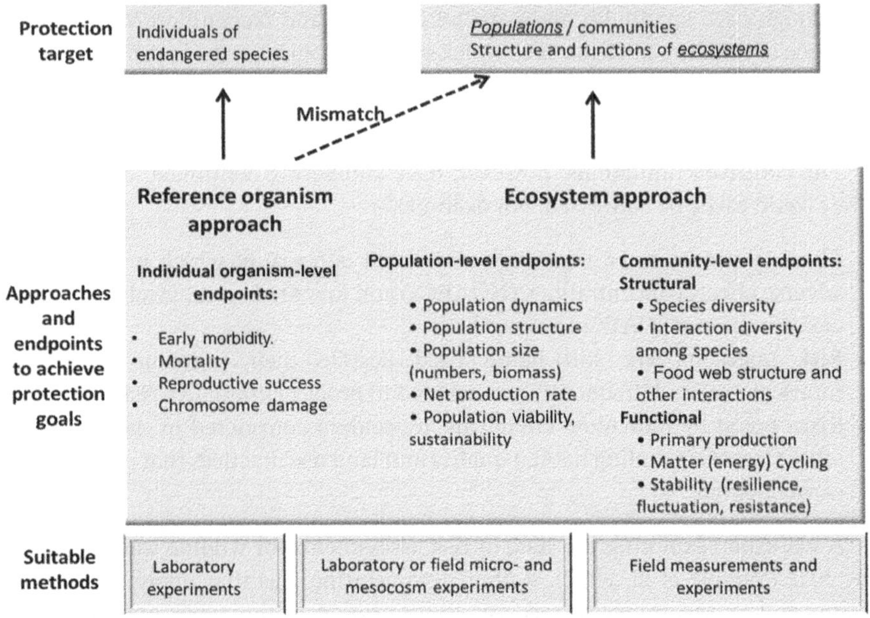

Figure 2.1 Mismatch between stated protection goals and the methods used to assess them. *Source:* Bréchignac et al. (2016) / with permission of Elsevier.

(World Population Clock: 8 Billion People [LIVE, 2023] – Worldometer [worldometers.info], accessed 29 May 2023) and is projected to reach 10 billion by 2050 (WDI – World's population will reach nearly 10 billion by 2050 [worldbank.org], accessed 29 May 2023). With humans already responsible for worsening climate crises, we need to consider that humans directly through massive alteration of landscape features and indirectly through climate change pose a dominant stressor effect in any setting we wish to focus on.

Not all the cases discussed in the following chapters explicitly refer to modifications of land and waterways nor do they all address the effects of climate. In those situations where human alterations and climate influences are not considered, it is likely that the risk characterization is incomplete and, where management options are described, adjustments will likely be needed within a few years as new information comes to light.

Biggs et al. (2015) described the areas of focus for achieving resiliency and sustainability within social-ecological systems. On the ecological side they argue for three features, namely, maintenance of diversity and redundancy, management of connectivity, and management of slow variables and feedback loops. On the social side, we should foster an understanding of social-ecological systems as complex adaptive systems, encourage learning and experimentation, broaden participation, and promote polycentric governance systems.

As we look at the pressing environmental challenges of our time (rapidly dwindling biodiversity; accelerated habitat fragmentation; the spread of plant and animal diseases, including those affecting humans; food, water, and energy security for human populations; and environmental justice), it is easy to become cynical about the future. We should recognize that none of these challenges will be addressed adequately so long as we continue to use poorly scoped risk assessments. We could dwell on the inadequacy of responses by scientists, politicians, and the public. But perhaps it will be more productive to search for ways to take the information gleaned from the cases described in this book and rededicate ourselves to find meaningful and workable solutions to the many environmental challenges we face in the dynamic and complex social-ecological system we live in.

References

ASTM (2023). E1963 –standard guide for conducting terrestrial plant toxicity tests.

Biggs, R., Schlüter, M., and Schoon, M.L. (2015). *Principles for Building Resilience: Sustaining Ecosystem Services in Social-Ecological Systems*, 290. Cambridge University Press.

Bradshaw, C., Kapustka, L., Barnthouse, L. et al. (2014). Using an ecosystem approach to complement protection schemes based on organism-level endpoints. *J. Environ. Radioact.* 136: 98–104.

Bréchignac, F., Oughton, D., Mays, C. et al. (2016). Addressing ecological effects of radiation on populations and ecosystems to improve protection of the environment against radiation: agreed statements from a consensus symposium. *J. Environ. Radioact.* 158–159: 21–29.

Hurlbert, S.H. (1984). Pseudoreplication and the design of ecological field experiments. *Ecol. Monogr.* 54: 187–211. https://doi.org/10.2307/1942661.

Kapustka, L. (2008). Limitations of the current practices used to perform ecological risk assessment. *Integr. Environ. Assess. Manag.* 4: 290–298.

Kapustka, L.A. and Landis, W.G. (2010). *Environmental Risk Assessment and Management from a Landscape Perspective*, 396. New York, NY: Wiley.

Ohno, Y. (2002). ICH guidelines – implementation of the 3Rs (refinement, reduction, and replacement): incorporating best scientific practices into the regulatory process. *ILAR J.* 43 (Suppl_1): PPS95–S98. https://doi.org/10.1093/ilar.43.Suppl_1.S95.

Sample, B., Johnson, M.S., Hull, R. et al. (2022). Key challenges and developments in wildlife ecological risk assessment: problem formulation. *Integr. Environ. Assess. Manag.* 20: 658–673. https://doi.org/10.1002/ieam.4710.

USEPA (1992). *Framework for Ecological Risk Assessment*. EPA/630/R-92/001. Washington, DC: U.S. Environmental Protection Agency.

Warren-Hicks, W., Parkhurst, B.R., and Baker, S.S. (ed.) (1989). *Ecological Assessments of Hazardous Waste Sites: A Field and Laboratory Reference*. 600/3-89/013. Washington, DC: United States Environmental Protection Agency.

3

Exposures and Effects

Cameron Irvine[1], *Mary A. Ottinger*[2], *and Lawrence A. Kapustka*[3]

[1] *Associate Environmental Scientist at Robertson-Bryan, Inc., Environment Department, Rancho Cordova, CA, USA*
[2] *Professor Emeritus and Research Professor, Department of Biology and Biochemistry, University of Houston, Houston, TX, USA*
[3] *LK Consultancy, Diamond Valley, AB, Canada*

Introduction

Early scientific experiments exposed organisms to nominal concentrations of a contaminant or chemical to study its lethality. From these simple toxicity studies, it was learned that "the dose makes the poison" (Paracelsus, the father of toxicology) – meaning that a substance produces a harmful effect only when the concentration to which the organism is exposed is sufficiently high and that there are no effects when exposures are sufficiently low (Langham and Kapur 2006). This classical toxicology adage is still relevant today; nevertheless, we have developed a much more detailed understanding of exposures and effects. We know that the effects caused by different compounds can vary and that the amount of a chemical that causes an effect differs among organisms and across different life stages. Modern toxicologists have gained insights into exposure pathways, how much of the toxicant is biologically available to the organism, sublethal effects (e.g., from molecular changes to growth and reproduction), and the mechanisms of effects within organisms. Scientists are also making strides to explain interactions among compounds, at population-level effects within and among species, and from a molecular to regional scale (Nowell et al. 2024).

Toxicity is complex. Some environmental toxicants (e.g., selenium and mercury) have multiple chemical forms and complex exposure pathways that differ with site-specific characteristics. For example, it is not the smallest organisms that are most sensitive to selenium, but instead it is the higher trophic-level egg-laying vertebrates (i.e., birds and fish) that are most sensitive. These aquatic-dependent wildlife bioaccumulate selenium through their diet, and the amount of selenium in their diet depends on the site-specific food web as well as the chemical form (i.e., oxidation state) of selenium present in surface water. Likewise, humans and aquatic-dependent wildlife are most sensitive to effects from mercury, specifically methylmercury, bioaccumulated through dietary exposure and affecting the nervous system. Thus, measuring contaminant concentrations in the exposure medium closest to the site of toxic action for the most sensitive organism provides the best predictor of adverse effects and is a reasonably accurate indication of risk to those organisms. Water column criteria exist for most metals, but they are being replaced by tissue-based water quality criterion elements for mercury and selenium. These media are fish tissues (whole fish or fillets) for mercury (USEPA 2001) and fish or bird eggs, whole fish, or fillets for selenium (USEPA 2021, 2024a). Investigating the drivers of effects to organisms where there are multiple stressors is much more complex than these single-contaminant models.

In this chapter, we review the strides taken to improve our understanding of environmental concentrations, organism exposures, effects, and how the tools for evaluating these conditions and processes can be integrated into models to assess the potential for adverse effects for both human health and wildlife. We also discuss the benefits, uncertainties, and limitations to which environmental toxicology and risk assessments are susceptible.

Analytical Tools

The contaminants and physical stressors affecting organisms in the environment can vary. Methods for measuring these chemicals and other factors, such as habitat loss or degradation (e.g., effects of invasive species, predation pressure, and increases in salinity and temperature associated with climate change), also vary due to local conditions and differences in measurement techniques. As such, diverse data may be produced that obscure relationships between stressors and related responses. Among measured environmental variables, chemical contaminants are one group of potential stressors to organisms in the environment that scientists have become quite good at measuring. Environmental samples can be collected with vetted and standardized practices to minimize contamination so that the measured concentration can be reported reliably (e.g., clean hands–dirty hands water sampling for low-level metals analysis using US Environmental Protection Agency [USEPA, 1996] Method 1669). Current analytical chemistry methods can be extremely sensitive at low concentrations that are often below that needed to trigger a physiological effect. Gas chromatography (GC), mass spectrometry (MS), high-performance liquid chromatography (HPLC), and combinations thereof are relatively common and accessible analytical laboratory instruments that can analyze environmental samples brought from the field. In recent decades, more specialized nuclear analytical techniques (e.g., synchrotron radiation and X-ray fluorescence) have improved sensitivity, accuracy, and resolution (Alam et al. 2022; Dinsley et al. 2022). Over the past few decades, these methods have improved from measuring parts per million (ppm) concentrations (mg/L in water), to parts per billion (ppb or μg/L in water), and the best of these currently can measure parts per trillion (ppt or ng/L in water) for some chemicals. For context, one ppt is analogous to a single drop of water in 20 Olympic-sized swimming pools (about 1 g of solid in a billion liters of water).

Assessing the meaning of detecting selected compounds at very low concentrations, both relative to the risk for adverse outcomes and the threshold concentrations that pose a risk to humans, wildlife, and ecosystems, represents an ongoing challenge for both old and new contaminants. Sensitive analytical methods are needed to determine concentrations that cause adverse effects for extremely toxic chemicals, some of which have been unintentionally released into the environment (e.g., mercury, PAHs, PCBs, pharmaceuticals, and personal care products), and others developed for intentional dispersion in the environment (e.g., pesticides). However, the capability to measure minute concentrations of specific compounds can have a double edge. Highly sensitive analyses exist, but are the results meaningful for adverse effects on wildlife and ecosystems? This is especially an issue for establishing thresholds for regulatory applications and actions. For example, detection of PCBs or PFAS compounds can be performed with high sensitivity, but the need for the lowest levels of analytical detection should be considered in light of the project goals, effort, and expense. Just because a chemical is found in the environment above a method detection limit does not mean that it poses a risk to organisms (including humans). Chemical analyses need to have detection limits that are at least at the level of criteria or benchmarks to which they are compared, if that is a goal or intent of sampling and analysis. Not all chemicals have regulatory thresholds, so the appropriate detection limit must be associated with an effect to be interpreted. In the procedures for establishing data quality objectives, the sensitivity of the analytical methods to be used should generally be an order of magnitude below the estimated threshold for biological or physiological response in the target organisms (USEPA 2006). If overly sensitive methods are used, then there is unnecessary cost in getting those data; if not sensitive enough, then useless data will be obtained.

The level of sensitivity needed for analytical measurements is easiest to understand for metals and other naturally occurring substances. Metals typically have effects associated with concentrations in the ppm range. Thus, detecting metals in the ppt range may not be necessary. On the other hand, manufactured organic compounds that were once thought to pose no risk of adverse effects have often been found to affect wildlife (e.g., DDT) or human health (e.g., PFAS) after studies investigate more sensitive endpoints or new receptors. For example, any concentration of PFOA or PFOS in surface water above detection limits (i.e., ~1 ppt) will exceed USEPA (2024b) draft water quality criteria for the protection of human health (i.e., 0.0009 and 0.06 ppt, respectively). Moreover, because these draft water quality criteria are below current detection limits, it is uncertain whether or not concentrations pose a risk to human health when not detected. It is notable that concentrations of these persistent organic pollutants in rainwater from many locations of the earth can exceed these extremely low benchmarks (Cousins et al. 2022).

Another example of the need for ever greater analytical sensitivity to meet effect levels is for pyrethroid insecticides, which are used for crop protection, flea and tick treatments in household pets, and residential pest control. Synthetic pyrethrins were derived from the chrysanthemum flower, and their use expanded in California, USA, among other places, during the 1990s as a replacement for organophosphate insecticides. Ambient concentrations in stormwater and sediment were soon found to cause acute mortality to aquatic invertebrates (Amweg et al. 2005, 2006). The freshwater amphipod, *Hyalella azteca,* is acutely sensitive to pyrethroids and concentrations ~1 ng/L caused significant mortality in 96-hour exposures (Weston and Lydy 2012). These concentrations were below most analytical method detection limits at the time of these studies, and more sensitive analytical methods were subsequently developed so that they are now detected at concentrations <1 ng/L.

Organisms (including humans) often encounter complex mixtures and, in fact, retain a body burden that will vary with the individual's exposures over time. As such, it is critical to ascertain if the observations field and laboratory studies are relevant to true and meaningful adverse outcomes and if these data are useful for regulatory applications (e.g., water quality criteria development). In the case of major chemical spills, there may be large-scale effects documented for several species and chemical exposure concentrations high enough to cause acute toxicity (i.e., mortality and other adverse effects associated with short-term exposures, relative to the organisms' lifespan). However, long-term low-level (i.e., chronic) consistent or intermittent exposures are likely more common for most humans and wildlife. Assessing the cause of adverse effects becomes even more challenging when changes are detected in individuals without associated exposure concentrations to understand if the change is associated with one or more specific toxicants. Some of the forthcoming discussions about the One Health and Exposome concepts and their frameworks can provide a roadmap to approach these questions and address current and emerging problems associated with environmental chemicals. Acquiring greater understanding of the relationship between exposures and effects on reproductive, immune, or other physiological effects is important for identifying both safe exposures and those that pose an unacceptable increase in risk of harm from nonlethal chronic exposures.

Toxicity Testing

Determining stressor effect levels for environmentally relevant ecological receptors is the next step in assessing the potential for adverse effects and understanding risks, after measuring the magnitude of a stressor. In their simplest form, laboratory toxicity studies (i.e., bioassays) consist of adding measured amounts of substances into clean water (or air, sediment, soil, or food items) and then introducing organisms of a single species into the test systems. Then, at specified intervals, the organism response is observed to document potential physiological effects (e.g., reduced survival, growth, reproduction, or other endpoints). Laboratory toxicity studies (i.e., bioassays) can either expose a test organism to an environmental sample or can isolate a specific compound and then test it at a range of doses, which may or may not be environmentally relevant, to determine concentrations associated with effects. These single compound tests (e.g., a pesticide in laboratory water) provide a controlled exposure, and the results of multiple exposure concentrations can be used to calculate effect concentrations (e.g., EC10, EC25, and ECx point estimates[1] of toxicity) from the concentration–response relationship. Testing an environmental sample has the benefit of integrating the effects of all chemical contaminants in the sample but has disadvantages in that interactions among them may not all be known and all chemicals present may not be measured. Serial dilutions are rarely conducted with environmental samples, and the results may not reflect exposures in nature if the addition of a clean dilution matrix affects the bioavailability and interactions among the mixture of chemicals.

The American Society for Testing and Materials (ASTM; now ASTM International) led the development of consensus-based standardized bioassay tests in the 1980s. These testing standards and guides described how to conduct a test and provided details pertaining to the culturing and care of test organisms. They listed the information that should be collected before, during, and after the test period, including the required test conditions and minimum test acceptability criteria that had to be met for a test result to be valid. ASTM guidance also cautioned about potential interferences that could influence the outcome of the test. These standards leveraged the knowledge of national experts developing methods to include suggested approaches to analyze, interpret, and report the test results.

Symposia were organized and panels of experts were assembled to design quality bioassay methods. Intense debates characterized the process, sometimes taking years, before consensus could be reached on every minute detail of tests

1 Ten percent effect concentration (EC10); 25% effect concentration (EC25). Point estimates of toxicity are calculated from a concentration–response relationship for any "point" (effect level) on a regression.

ranging from handling and culturing conditions, size of test units, number of replicates, duration of tests, endpoints to measure, verification of exposure conditions for the duration of the test, appropriate study design, appropriate statistical methods, and so on. For example, 96-hour tests were designed for setup on a Monday and takedown on Friday. Longer tests would require technicians to work weekends, substantially increasing test costs. Choosing the most important endpoints was among the topics that animated the symposia and panels. One line of argument was that mortality endpoints were the most important. The counterargument, from an ecological perspective, was that reduced fitness had greater consequence than death because the compromised organisms would continue to access and consume resources, which would translate to lowered fecundity and thus have an adverse population effect. Eventually, consensus was reached to record multiple endpoints at multiple time steps during the tests. Thus, various measures of metabolic, physiologic, morphologic, developmental, and even behavioral endpoints were adopted in addition to the basic survival, growth, and reproduction endpoints characterizing early toxicity testing. Concurrently with the development of ASTM standard toxicity testing methods, many of the same national experts writing ASTM standards also developed standard bioassays for regulatory testing. USEPA (2002a,b) whole effluent toxicity (WET) tests and sediment toxicity tests (USEPA 2000) mimic their ASTM methods (e.g., ASTM 2020, 2022a, 2023) and were developed to provide a consistent and repeatable measure of toxicity with test organisms that reflect a range of valued ecosystem components.

Starting with toxicity tests focused on freshwater organisms (algae, zooplankton, invertebrates, and fish), the suite of tests expanded to address marine environments, sediments, soils, terrestrial plants, birds, and mammals. Regulatory guidelines, in recognition of the difficulty of relating toxicity test results to ecological conditions, recommended using batteries of test species representing different life-history strategies (Data Requirements for Pesticide Registration, USEPA) selected in part from ASTM methods (see ASTM Biological Effects and Environmental Fate[2]). Standardized toxicity test methods have been developed for aquatic plants (producers), amphipods, annelids, mollusks, echinoderms, crustaceans, shellfish, finfish, birds, and mammals.

Terrestrial test methods lagged those for aquatic systems, except for Safety Testing under Federal Insecticide, Fungicide, and Rodenticide Act (FIFRA) in the United States, which focused on agronomic plant taxa, beneficial insects, and avian species. Initially, under the Comprehensive Environmental Response, Compensation, and Liability Act (CERCLA; aka Superfund) program in the United States, terrestrial systems were characterized using eluates of soil

2 https://www.astm.org/jurisdiction-e5047.

samples to evaluate the response of aquatic test species (primarily *Raphidocelis subcapitata* [previously subcapitata *Pseudokirchneriella subcapitata* and *Selenastrum capricornutum*]). As Superfund gained greater societal awareness, pleas for more relevant assessments and test methods led to development of toxicity tests in soils. This involved using samples from contaminated sites as well as artificial soils that were intended to establish commonality among testing laboratories. To address questions of ecological relevance of test species, ASTM established criteria for qualifying any plant species of interest as a test species (ASTM 2022b). Toxicity test methods were developed for earthworms, nematodes, and springtails in soils (see ASTM Biological Effects and Environmental Fate[3]). Tests for birds and mammals, adapted from the FIFRA test guidelines, were largely restricted to the northern bobwhite quail, mallard duck, and various strains of mice. There are thousands of chemicals in the environment, and it has taken great effort to test many of these compounds with just some of the standard test species under standard test conditions. It would be an immense effort to test each chemical with the standard test species and under the range of environmental conditions that can affect the bioavailability of a toxicant (e.g., various pH, temperature, salinity, hardness, and alkalinity). Tests that have been conducted are typically reported in scientific literature, and many have been compiled into public databases (e.g., the USEPA EcoTox database[4]).

Through all of this, awareness grew as research began to show that single species tests, although useful in setting certain thresholds pertaining to single chemical stressors, were poor at predicting effects within complex ecological settings. This was demonstrated in a thesis study conducted at Oregon State University, Corvallis, Oregon (Pfleeger and Zobel 1995) in which Atrazine, 2,4-D, and malathion were administered at one-tenth the label concentrations to plots that had plant communities established using soil containing the seed bank from an annually plowed field that had no pesticide application for over 10 years. Productivity measured for the entire community, decreased with atrazine and 2,4-D application and species abundance was decreased in all treated plots, relative to controls. Results of this type and magnitude could not be predicted from single-species phytotoxicity tests.

Field studies conducted under FIFRA in the 1990s were expensive and had low statistical power, to the point of not being useful for pesticide registration decisions. The reason they produced little useful information relates to the number of replicate test areas, the low number of birds occupying the fields, and the relatively high variability in terms of population numbers among

---

3 https://www.astm.org/jurisdiction-e5047.

4 https://cfpub.epa.gov/ecotox/; https://www.epa.gov/comptox-tools/downloadable-computational-toxicology-data.

the test fields. More informative studies arose from incident reports after pesticides were authorized for use.

Laboratory testing has value in allowing control over variables that can be helpful for understanding the mechanisms of exposures and toxic effects. However, despite the consistent methods and repeatability of standardized laboratory bioassays, they were criticized as not being representative of real-world conditions. Likewise, field experiments were challenged as being plagued by variability and pseudoreplication (Hurlbert 1984). Mesocosm studies overcome these criticisms by creating small ecosystems where environmental testing can be conducted. They were developed to provide the benefits of both laboratory and field designs. Standard test species and endpoints can be observed while temperatures, sunlight, organisms, and exposure concentrations are controlled and other natural environmental conditions can be largely maintained (e.g., Fordham 2015; Hasenbein et al. 2016). Some of the critiques of mesocosms and field studies were that they often produced results that were not predicted from single-species, organism-level tests. Why this was a surprise to scientists is attested to the general lack of understanding regarding the complexity of ecological systems and the varying sensitivities of different organisms. Study results demonstrate the environmental relevance of mesocosm testing where exposure conditions resemble those in nature more than laboratory tests. Indeed, emergent properties that arise as one moves through the hierarchy of organism to population to community to ecological system should alert us that different results should be expected.

Integrating multiple lines of evidence is another approach for environmental assessments to determine if valued ecosystem resources are being adversely affected. Comparing the results of toxicity testing where samples are brought back to the lab, with *in situ* toxicity testing where test organisms are exposed to site media in the field, can create a powerful weight of evidence when combined with chemical analyses results and a thoughtful conceptual model of exposure pathways (Burton et al. 2005). Not relying on any single type of test requires a reasonable understanding of the site conditions, chemical fate and transport, and environmental responses to explain what might appear on the surface to be confounding results. For example, by measuring pore water upwelling and downwelling, Greenburg et al. (2002) were able to determine that surface water–pore water interactions can have either protective or harmful effects on organisms in the environment and explain differences between *in situ* field exposure and tests conducted in the laboratory. Clean surface water downwelling can protect sediment dwelling organisms in contaminated sediment that causes mortality to freshwater invertebrates when tested in the laboratory. Conversely, contaminated groundwater upwelling can cause toxicity in sediments that are not toxic in the laboratory. Various other comprehensive approaches have been helpful for assessing overall environmental quality, but they consider contaminant mixtures and multiple types of stressors to various degrees and as needed on a case-by-case basis. A challenge when conducting these field studies is identifying an appropriate reference site, as there are rarely locations with potential toxicant exposures with identical locations absent those exposures. Regulatory guidance documents have been developed to address this issue (e.g., USEPA 2002c; ITRC 2021; ASTM 2024).

An increasing number of studies are assessing the effects of mixtures of contaminants (e.g., Harrahy and Clements 1997; Callinan-Hoffman et al. 2012; Fettweis et al. 2023). Binary (two-compound) and ternary (three-compound) tests represent initial laboratory efforts to determine interactions among the multitude of chemicals to which organisms can be exposed. Nevertheless, these tests are relatively more intensive (i.e., higher labor effort and cost) than assessing single contaminant toxicity due to the combinations of constituents requiring a greater number of treatments. Such studies have determined that contaminant effects can be additive, less than additive, or synergistic. Additive effects occur when toxicants have the same mode of action and the total effect is equal to the sum of the effects from individual toxicants. This is referred to as concentration addition (CA). Chemicals can demonstrate independent action (IA) when the mode of toxic action differs so that toxicity is less than the sum of individual toxicants (Escher et al. 2020). Synergy occurs when the effect is greater than the sum of the individual toxicants. These interactions among toxicants add a level of complexity to environmental toxicology that we have only recently begun to understand.

Predictive Toxicology

Studies investigating how toxicity changes under different environmental conditions have identified important variables that modify toxicity. Incorporating these relationships into models has improved the predictability of toxicity for several metals and organic contaminants. It has long been known that the dissolved fraction of cationic metals is bioavailable and not the total concentration, as ions can more easily interact with organisms than metals bound to particulates. Further bioavailability refinements for metals in freshwater systems have included adjustments for hardness because higher hardness contains more chemical ligands to bind the metals and reduce toxicity. Thus, USEPA National Recommended Water Quality Criteria for several metals (i.e., cadmium, chromium, lead, nickel, silver, and zinc) are based on dissolved concentrations (i.e., filtered) and have included hardness adjustments to account for site-specific conditions.

These water quality criteria adjusted for hardness are still in use by many jurisdictions. Since then, studies have shown that other chemical properties also affect the bioavailability of cationic metals.

The Biotic Ligand Model (BLM [5]) was developed in the 1990s–2000s to calculate the bioavailable fraction of a metal after accounting for interactions with other major ligands (i.e., dissolved organic carbon, calcium, chloride, magnesium, potassium, sodium, sulfate, bicarbonate, carbonate, carbonic acid) and exposure conditions (i.e., pH and temperature) to predict acute and chronic toxicity to aquatic organisms (Paquin et al. 2002; Ryan et al. 2004). BLM models are available for aluminum, cadmium, cobalt, copper, lead, nickel, and zinc in freshwater, and for copper, nickel, and zinc in saltwater (Santore et al. 2002; Arnold et al. 2005; DeForest and Van Genderen 2012; DeForest et al. 2017; Santore et al. 2018). The USEPA (2007) adopted the BLM as the basis for determining national recommended ambient water quality criteria for copper, and it is being considered for regulatory use in California as site-specific objectives for copper and zinc (Los Angeles Regional Water Quality Control Board 2019). Expanding on the BLM for single metals, a multimetal biotic ligand model (mBLM) was developed for predicting bioavailability and toxicity of metal mixtures to aquatic organisms (Santore and Ryan 2015). This model is consistent with observations where the toxicity of multiple metals is either additive or less than additive. Metal toxicity is additive because they have the same mode of toxic action but can be less than additive when the metals compete for a limited number of binding sites where they interact with the organism. Similar modeling approaches, such as multiple linear regression (MLR), have also been developed that account for metal interactions with other ions (e.g., DOC and pH) to predict toxicity in surface water and determine concentrations that are protective of aquatic organisms in water with differing chemical compositions (Brix et al. 2023). The BLM has also been applied to describe metal toxicity in sediment (DiToro et al. 2005).

There are several approaches for predicting toxicity from contaminants in sediment. Organic compounds and cationic metals bind to organic carbon in soils and sediment such that normalizing pollutant concentrations for organic carbon produces predictable toxicity patterns. Di Toro et al. (1991) described this organic carbon normalization in sediments. Such an equilibrium partitioning (EqP) approach was chosen by the USEPA (2008) as a basis for developing sediment quality criteria for organic compounds because it accounts for the bioavailable fraction of numerous chemicals in different sediments. Procedures for determining equilibrium partitioning–based sediment benchmarks (ESBs) for mixtures of several metals (i.e., cadmium, copper, lead, nickel, silver, and zinc) that are protective of freshwater benthic organisms are based on the determination of

5 https://www.windwardenv.com/biotic-ligand-model/.

bioavailable metal concentrations in porewater (USEPA 2005). Metals in sediment are potentially bioavailable when concentrations of simultaneously extracted metals (SEMs) exceed the concentration of acid volatile sulfides (AVSs). Sediment toxicity is predicted when these excess metals, divided by the fraction of organic carbon (f_{OC}) ($[\Sigma SEM - AVS]/f_{OC} > 0$) are greater than their effect levels in water. Evaluation and refinements to these models indicate that OC normalization of excess SEM should not be considered for sediments with an OC concentration <1% and is questionable for sediments with an OC concentration of 1–4% to minimize the likelihood of incorrectly identifying nontoxic samples as toxic (DeForest et al. 2021).

Pyrethroid insecticides provide an example of how it can take several decades to learn enough about a new chemical so that its toxicity can be predicted. After a decade of use, studies in the Central Valley of California were finding substantial toxicity in stormwater and sediment samples associated with pyrethroids (Amweg et al. 2005, 2006; Weston and Lydy 2012). Coincidentally, the test organism used in these studies, *H. azteca*, a freshwater epibenthic amphipod, is one of the most sensitive known organisms to pyrethroids. Concentrations in the single-digit ppt range can immobilize or cause mortality within 96 hours. Synthetic pyrethroids, of which there are several (e.g., bifenthrin, cypermethrin, esfenvalerate, lambda-cyhalothrin, deltamethrin, and permethrin), have the same mode of toxic action, and their toxicity is additive. Pyrethroids are also highly hydrophobic and bind strongly to sediment and organic carbon; thus, their toxicity in surface water is greatly influenced by the dissolved fraction. This information led to changes in use but also to water quality goals being developed in the Central Valley of California above which discharges are prohibited. These goals consider the dissolved fraction of the six pyrethroids listed earlier and the sum of their toxicity (Central Valley Water Board 2019). Data may be starting to show a decline in the use of bifenthrin and cypermethrin in parts of California (Budd et al. 2020). However, deltamethrin concentrations may be increasing. This change in the use of these chemicals may only be a step in the cycle of chemical substitutions where new chemicals are developed to take the place of old ones found to cause unintended adverse effects. Imidacloprid, another relatively new neonicotinoid pesticide with links to effects in pollinators (e.g., bees), has been banned for nonagricultural outdoor uses in California starting in 2025.

Increasingly Sensitive Endpoints and Expanding Knowledge

In recent years, testing endpoints has expanded beyond survival, growth, and reproduction to include more subtle effects. Behaviors such as swimming and avoidance responses of larval fish and invertebrates can be recorded and have shown alterations with exposure to environmental chemicals (Cantu et al. 2023). Video observations are being analyzed using software to discern patterns without human bias and countless researcher hours. These behaviors can be captured for a large number of organisms at once (e.g., larval zebrafish or larvae frog in 96-well plates) to allow many individual exposures (e.g., Geach et al. 2012; Shen et al. 2020). This is an example of high-throughput screening that can assess the toxicity of many different samples or chemicals in a relatively short period of time. The USEPA Toxicity Forecasting (ToxCast[6]) tool makes available to the public in vitro and high-throughput screening assay data for almost 10,000 substances.

Nonlethal effects of environmental chemicals can include endocrine disruption (for reviews, see Ottinger and Dean 2022; Ottinger and Quinn 2024). There are varied physiological and endocrine effects of these endocrine-disrupting chemicals (EDCs), which can include reproductive, metabolic, stress response, thyroid and growth, and immunocompetence responses. Such nonlethal effects can exert lifelong effects that may limit health, longevity, and productivity of individuals, with potential impacts on populations (Ottinger and Dean 2022; Grace et al. 2024).

The study of molecular endpoints (e.g., genomics, transcriptomics, proteomics, and metabolomics) or "omics" is an entire field of science providing insight into the inner workings of organisms. Measuring upregulation and downregulation of gene expression (i.e., genomics), gene transcription (i.e., transcriptomics), or protein production (i.e., proteomics) can result in the analysis of hundreds of endpoints at once and be conducted with material collected by nonlethal sampling. Moreover, the range of molecular endpoints can be customized for chemical classes and to optimize the predicted responses indicative of adverse physiological effects. The EcoToxChip is an example of this where an environmental sample can be evaluated with a single "chip" (e.g., a 384-well quantitative polymerase chain reaction [qPCR] array) that assays hundreds of genes associated with multiple species (Basu et al. 2019; Crump et al. 2023; Mittal et al. 2023). Adverse Outcomes Pathways (AOPs) are the conceptual framework linking molecular and hormonal measures to apical effects in the whole organism (e.g., survival, growth, and reproduction) (Villeneuve et al. 2014a,b).

An exhaustive list of the tools used by toxicologists, environmental scientists, and managers has not been presented here. Only a few examples are discussed to highlight the processes that need to be considered when evaluating effects from multiple stressors. There is still very much to learn. Basic science (e.g., concentration–response relationships for chemicals and species) is needed to inform complex integrative tools, but we are experiencing a renaissance in

---

6 https://www.epa.gov/comptox-tools/toxicity-forecasting-toxcast.

our understanding of toxicology and big data (i.e., open access databases of gene regulations such as EPA AOP-DB,[7] GEO,[8] JASPAR,[9] and others), AOPs, and quantitative structure–activity relationships (QSARs) – methods for predicting the properties and actions of compounds based on their chemical structure. Assessing all of this available knowledge will enable more predictive models and earlier intervention of the adverse effects of environmental chemicals on wildlife and humans (Head et al. 2024). One can only imagine how leveraging the power of artificial intelligence with all of the available resources will further expand our knowledge horizons and address the complexity associated with understanding effects from multiple stressors.

The Exposome and One Health

The Exposome is a concept that primarily focuses on the individual organism. More specifically, the Exposome considers external and internal exposures from an individual's environment, which includes chemical pollutants, shelter and living conditions (i.e., habitat), noise, and lifestyle/behaviors, such as diet, exercise, and sleep, as they affect the health of an individual throughout their lifespan (Wild 2012). This concept is intriguing in that the Exposome attempts to integrate quantitative and qualitative environmental factors that affect an individual over time. The Exposome becomes important in understanding the response of an individual. It is also important in developing lifetime measures of exposure and their effects on an individual. Currently, the Exposome Concept is being used to address human health with a focus on the cumulative effects of environmental chemical exposures on the individual (Ottinger and Geiselman 2023).

There are many aspects that influence an individual's Exposome, including genetics, lifestyle, disease, environment, and community. The Exposome of an individual can be assessed at any point in time or at various stages of their lifetime, with the notion that many of the effects of environmental factors will be cumulative. Furthermore, the power of the Exposome is that there are now capabilities to measure physiological endpoints affected by these exposures. These endpoints include molecular, genomic, and metabolic measures, as well as those of the whole organism (e.g., survival, growth, and reproduction), and disease states such as cancer. Together, these measurement endpoints, termed Exposomics, can provide targeted biomarkers (when linked to apical endpoints) and responsive endpoints (e.g., survival, growth, and reproduction) for exposures and environmental stressors affecting an individual. When the

---

7 https://www.epa.gov/healthresearch/adverse-outcome-pathway-database-aop-db.
8 https://www.ncbi.nlm.nih.gov/geo/.
9 http://jaspar.genereg.net/.

relationship between stressors and health effects are known, Exposomics can provide data for modeling adverse effects. The measurement endpoints under the umbrella of Exposomics can include AOPs, among other stressor–response relationships. The Exposome approach can also be applied to wildlife and human health. In fact, as discussed earlier, the application of Exposomics to wildlife will provide insights and predictive indicators of health.

The One Health concept is defined by the US Centers for Disease Control and Prevention as "a collaborative, multisectoral, and transdisciplinary approach – working at the local, regional, national, and global levels – with the goal of achieving optimal health outcomes recognizing the interconnection between people, animals, plants, and their shared environment" (https://www.cdc.gov/onehealth/about/?CDC_AAref_Val=https://www.cdc.gov/onehealth/basics/index.html). Although there have been cultural and societal practices that acknowledge and respect the role of the environment and nature for healthy humans, there has not been a structured conceptual framework for the comprehensive evaluations of exposures and effects until more recently. A primary catalyst for the One Health concept and its associated framework arose from the recognition of the risk of disease transmission from wildlife to humans, and there is still a major emphasis on veterinary and immune/disease aspects of One Health (for detailed information, see Ottinger and Geiselman 2023). It has only been more recently that the One Health concept has become more encompassing to include many factors, including zoonotic disease that influences the resilience and quality of an ecosystem and its inhabitants. As such, the One Health concept can be exceedingly complex as it considers the many factors that interact and affect the resilience of ecosystems (Figure 3.1). The One Health Conceptual Framework takes advantage of these complexities to bring together an approach for organizing a response to an event. For example, an issue that stems from a catastrophic event, as was the case for the Deep-Water Horizon Oil Spill in the Gulf of Mexico, had physical, chemical, and environmental implications for humans and wildlife, with potential short- and long-term ramifications for ecosystems in the region of concern. Considering the approaches used to respond to an event, such as this oil spill, a One Health framework could have provided a structured approach to assess the situation, followed by the development of a stepwise plan to mitigate, restore, and assess the outcomes. The power of this structured approach is the involvement of research and public entities, with engagement of stakeholders in the affected communities.

There are a number of national, international, and academic One Health programs focused on gaining a better understanding of the dynamics related to health and the environment. Some of these programs primarily focus on disease transmission – detection, mitigation, and biocontrol. Although the

Ingredients for human health	Effects of current and mounting challenges				
	Population growth	Novel disease	Anthro pollution	Climate change	Social/economic disparities
Clean water	→	→	↓	↓	↓
Adequate nutrition	→	↓	↓	→	↓
Clean air	↓	→	↓	↓	↓
Appropriate shelter	→	↓	↓	↓	↓
Medical care	→	↓	↓	↓	↓
Nature/recreation	↓	→	↓	→	↓
Human community	→	↓	↓	↓	→
Toxic substances	↓	→	↓	↓	↓
Infectious diseases	↓	→	↓	↓	↓
Trauma	↓	→	N/A	↓	↓
Lifestyle	N/A	→	↓	→	↓

Figure 3.1 The essential ingredients that affect individual human health (Exposome) are interactive and involve the environment (One Health). *Source:* Ottinger and Geiselman (2023) / with permission of Elsevier. Shaded rows indicate ingredients that should be avoided while unshaded rows are those needed for good health (→ = combinations of positive and negative effects; ↓ = negative effects; ↑ = positive effects; n/a = not applicable). Blue headings are for challenges at least partly addressed by the Exposome framework; headings with green shading are for challenges that have been addressed by both One Health and the Exposome; headings without shading identify challenges that neither framework adequately addresses.

Exposome concept primarily focuses on individual exposures at specific stages in the lifespan and over the lifetime, there are extremely valuable insights from the responses of individuals to chemicals and environmental stressors over the lifespan. The issue of health span is one concerning all organisms. The use of Exposomics provides an opportunity to develop risk assessment approaches that are tuned to specific scenarios. Finally, there are some examples in which both the One Health and the Exposome have played a part in attacking specific issues, such as the aforementioned example of the Deep-Water Horizon Oil Spill. In addition, global problems arising from the use of cook stoves, which give rise to toxic airborne chemicals, and the spread of disease throughout communities all provide examples of how these concepts can be applied, even if the application is not intentional as using a One Health or Exposome framework. The integration of the Exposome Framework, Exposomics, and the One Health Framework provides structured approaches to characterize the relationship between multiple stressors and environmental variables and develop interventions to reduce adverse effects.

References

Alam, M.A., Tiwari, M.K., Trivedi, A. et al. (2022). Improvement of limit of detection sensitivities in the parts per billion range using conventional geometry synchrotron radiation excited EDXRF measurements. *J. Anal. At. Spectrom.* 3: 575–583. https://doi.org/10.1039/D2JA00016D.

Amweg, E.L., Weston, D.P., and Ureda, N.M. (2005). Use and toxicity of pyrethroid pesticides in the Central Valley, California, USA. *Environ. Toxicol. Chem.* 24 (4): 966–972. https://doi.org/10.1897/04-146R1.1.

Amweg, E.L., Weston, D.P., You, J., and Lydy, M.J. (2006). Pyrethroid insecticides and sediment toxicity in urban creeks from California and Tennessee. *Environ. Sci. Technol.* 40 (5): 1700–1706. https://pubs.acs.org/doi/10.1021/es051407c.

Arnold, W.R., Santore, R.C., and Cotsifas, J.S. (2005). Predicting copper toxicity in estuarine and marine waters using the biotic ligand model. *Mar. Pollut. Bull.* 50: 1634–1640.

ASTM International (2020). Standard test method for measuring sed tox freshwater invertebrates. E1706-20. https://www.astm.org/membership-participation/technical-committees/committee-e50/subcommittee-e50/jurisdiction-e5047.

ASTM International (2022a). Standard guide for conducting three-brood, renewal toxicity tests with *Ceriodaphnia dubia*. E1295-22. https://www.astm.org/membership-participation/technical-committees/committee-e50/subcommittee-e50/jurisdiction-e5047.

ASTM International (2022b). Standard guide for conducting terrestrial plant toxicity tests. E1963-22. https://www.astm.org/membership-participation/technical-committees/committee-e50/subcommittee-e50/jurisdiction-e5047.

ASTM International (2023). Standard guide for conducting acute toxicity tests on aqueous ambient samples and effluents with fish, inverts, amphibians. E1192-23. https://www.astm.org/membership-participation/technical-committees/committee-e50/subcommittee-e50/jurisdiction-e5047.

ASTM International (2024). Standard guide for selection of background reference areas for determination of representative sediment background concentrations. E3344-24. https://www.astm.org/membership-participation/technical-committees/committee-e50/subcommittee-e50/jurisdiction-e5004.

Basu, N., Crump, D., Head, J. et al. (2019). EcoToxChip: a next-generation toxicogenomics tool for chemical prioritization and environmental management. *Environ. Toxicol. Chem.* 38 (2): 279–288. https://doi.org/10.1002/etc.4309.

Brix, K.V., Tear, L., DeForest, D.K., and Adams, W.J. (2023). Development of multiple linear regression models for predicting chronic iron toxicity to aquatic organisms. *Environ. Toxicol. Chem.* 42 (6): 1386–1400. https://doi.org/10.1002/etc.5623.

Budd, R., Wang, D., Ensminger, M., and Phillips, B. (2020). An evaluation of temporal and spatial trends of pyrethroid concentrations in California surface waters. *Sci. Total Environ.* 718: 137402. https://doi.org/10.1016/j.scitotenv.2020.137402.

Burton, G.A. Jr., Greenberg, M.S., Rowland, C.D. et al. (2005). *In situ* exposures using caged organisms: a multi-compartment approach to detect aquatic toxicity and bioaccumulation. *Environ. Pollut.* 134: 133–144. https://doi.org/10.1016/j.envpol.2004.07.008.

Callinan-Hoffmann, K., Deanovic, L.A., Stillway, M., and Teh, S.J. (2012). *The Toxicity and Interactions Among Common Aquatic Contaminants in Binary Mixtures*, 63. Sacramento, CA: State Water Resources Control Board.

Cantu, E., Rivera, M., Lacy, B., and Rahman, M.S. (2023). Effects of short-term exposure to a pesticide mixture on free-swimming behavior in goldfish, *Carassius auratus*. *J. Hazard. Mater. Adv.* 11: 100350. https://doi.org/10.1016/j.hazadv.2023.100350.

Central Valley Regional Water Quality Control Board (2019, February). *Board Water Quality Control Plan (Basin Plan) for the Central Valley Regional Water Quality Control Board Central Valley Region*, 5e. https://www.waterboards.ca.gov/centralvalley/water_issues/basin_plans/sacsjr_201902.pdf.

Cousins, I.T., Johansson, J.H., Salter, M.E. et al. (2022). Outside the safe operating space of a new planetary boundary for per- and polyfluoroalkyl substances (PFAS). *Environ. Sci. Technol.* 2022 (56,16): 11172–11179. https://pubs.acs.org/doi/10.1021/acs.est.2c02765.

Crump, D., Hickey, G., Boulanger, E. et al. (2023). Development and initial testing of EcoToxChip, a novel toxicogenomics tool for environmental management and chemical risk assessment. *Environ. Toxicol. Chem.* 42: 1763–1771. https://doi.org/10.1002/etc.5676.

DeForest, D.K. and Van Genderen, E.J. (2012). Application of USEPA guidelines in a bioavailability-based assessment of ambient water quality criteria for zinc in freshwater. *Environ. Toxicol. Chem.* 31: 1264–1272.

DeForest, D.K., Santore, R.C., Ryan, A.C. et al. (2017). Development of biotic ligand model (BLM)-based freshwater aquatic life criteria for lead following U.S. EPA guidelines. *Environ. Toxicol. Chem.* 36 (11): 2965–2973.

DeForest, D.K., Toll, J.E., Judd, N.L. et al. (2021). Sediment toxicity data and excess simultaneously extracted metals from field-collected samples: comparison to United States Environmental Protection Agency benchmarks. *Integ. Environ. Assess. Manage.* 18: 184–186. https://doi.org/10.1002/ieam.4462.

Di Toro, D.M., Zarba, C.S., Hansen, D.J. et al. (1991). Technical basis for establishing sediment quality criteria for nonionic organic chemicals using equilibrium partitioning. *Environ. Toxicol. Chem.* 10: 1541–1583.

Dinsley, J.M., Davies, H.S., Gomez-Gonzalez, M.A. et al. (2022). The value of synchrotron radiation X-ray techniques to explore microscale chemistry for ecology and evolution research. *Ecosphere* 13 (12): e4313. https://doi.org/10.1002/ecs2.4312.

DiToro, D.M., McGrath, J.A., Hansen, D.J. et al. (2005). Predicting sediment metal toxicity using a sediment biotic ligand model: methodology and initial application. *Environ. Toxicol. Chem.* 24 (10): 2410–2427.

Escher, B., Braun, G., and Zarfl, C. (2020). Exploring the concepts of concentration addition and independent action using a linear low-effect mixture model. *Environ. Toxicol. Chem.* 39 (12): 2552–2559. https://doi.org/10.1002/etc.4868.

Fettweis, A., Hansul, S., De Schamphelaere, K., and Smolders, E. (2023). Metal mixture toxicity of Ni, Cu, and Zn in freshwater algal communities and the correlation of single-species sensitivities among single metals: a comparative analysis. *Environ. Toxicol. Chem.* 42: 2666–2683.

Fordham, D.A. (2015). Mesocosms reveal ecological surprises from climate change. *PLoS Biol.* 13: e1002323. https://doi.org/10.1371/journal.pbio.1002323.

Geach, T.J., Stemple, D.L., and Zimmerman, L.B. (2012). Genetic analysis of *Xenopus tropicalis*. *Methods Mol. Biol.* 917: 69–110. https://doi.org/10.1007/978-1-61779-992-1_5.

Grace, J.J., Duran, E., Ottinger, M.A., and Maness, T. (2024). Sublethal effects of early-life exposure to common and emerging contaminants in birds. *Curr. Res. Toxicol.* 1: 100190.

Greenburg, M.S., Burton, G.A. Jr., and Rowland, C.D. (2002). Optimizing interpretation of *in situ* effects of riverine pollutants: impact of upwelling and downwelling. *Environ. Toxicol. Chem.* 21: 289–297. https://doi.org/10.1002/etc.5620210210.

Harrahy, E.A. and Clements, W.H. (1997). Toxicity and bioaccumulation of a mixture of heavy metals in *Chironomus tentans* (Diptera: Chironomidae) in synthetic sediment. *Environ. Toxicol. Chem.* 16 (2): 317–327. https://doi.org/10.1002/etc.5620160230.

Hasenbein, S., Lawler, S.P., Geist, J., and Connon, R.E. (2016). A long-term assessment of pesticide mixture effects on aquatic invertebrate communities. *Environ. Toxicol. Chem.* 35 (1): 218–232. https://doi.org/10.1002/etc.3187.

Head, J., Ewald, J.D., and Basu, N. (2024). The DIKW of transcriptomics in ecotoxicology: extracting information, knowledge, and wisdom from big data. *Environ. Toxicol. Chem.* https://doi.org/10.1002/etc.5954.

Hurlbert, S.H. (1984). Pseudoreplication and the design of ecological field experiments. *Ecol. Mongr.* 54: 187–211. https://doi.org/10.2307/1942661.

Interstate Technology Regulatory Council (ITRC) (2021, December). Soil background and risk assessment. https://sbr-1.itrcweb.org/wp-content/uploads/2022/02/SBR-Guidance-Document-Final-1-28-2022-Web-1.pdf.

Langham, L.J. and Kapur, B.M. (2006). Toxicology: then and now. *Clin. Biochem.* 39 (5): 498–510. https://doi.org/10.1016/j.clinbiochem.2006.03.004.

Los Angeles Regional Water Quality Control Board (2019). Preliminary implementation considerations for application of BLM-derived copper criteria in the Los Angeles region. Draft Report. 8 July 2019.

Mittal, K., Ewald, J., Crump, D. et al. (2023). Comparing transcriptomic responses to chemicals across six species using the EcoToxChip RNASeq database. *Environ. Toxicol. Chem.* https://doi.org/10.1002/etc.5803.

Nowell, L.H., Moran, P.W., Waite, I.R. et al. (2024). Multiple lines of evidence point to pesticides as stressors affecting invertebrate communities in small streams in five United States regions. *Sci. Total Environ.* 915 (10): 169634. https://doi.org/10.1016/j.scitotenv.2023.169634.

Ottinger, M.A. and Dean, K.D.M. (2022). Actions of toxicants and endocrine disrupting chemicals in birds. In: *Sturkie's Avian Physiology*, 2e (ed. C.G. Scanes), 1353–1384. Cambridge, MA: Elsevier Press.

Ottinger, M.A. and Geiselman, C. (2023). *One Health Meets the Exposome: Human, Wildlife, and Ecosystem Health*, 271. Academic Press.

Ottinger, M.A. and Quinn, M.J. Jr. (2024). Endocrine disruption in birds. In: *Hormones and Reproduction in Vertebrates*, vol. 4 (ed. D.O. Norris and K.H. Lopez). Oxford and San Diego, CA: Elsevier Press. pp. 263–299.

Paquin, P.R., Gorsuch, J.W., Apte, S. et al. (2002). The biotic ligand model: a historical overview. *Comp. Biochem. Physiol. Part C.* 133: 3–35.

Pfleeger, T. and Zobel, D. (1995). Organic pesticide modification of species interactions in annual plant communities. *Ecotoxicology* 4: 15–37. https://pubmed.ncbi.nlm.nih.gov/24197547/.

Ryan, A.C., Van Genderen, E.J., Tomasso, J.R. et al. (2004). Influence of natural organic matter source on copper toxicity to larval fathead minnows (*Pimephales promelas*): implications for the biotic ligand model. *Environ. Toxicol. Chem.* 23: 1567–1574.

Santore, R.C. and Ryan, A.C. (2015). Development and application of a multimetal multibiotic ligand model for assessing aquatic toxicity of metal mixtures. *Environ. Toxicol. Chem.* 34 (4): 777–787. https://doi.org/10.1002/etc.2869.

Santore, R.C., Mathew, R., Paquin, P.R., and Di Toro, D.M. (2002). Application of the biotic ligand model to predicting zinc toxicity to rainbow trout, fathead minnow, and *Daphnia magna*. *Aquat. Toxicol.* 133: 271–285.

Santore, R.C., Ryan, A.C., Kroglund, F. et al. (2018). Development and application of a biotic ligand model for predicting the chronic toxicity of dissolved and precipitated aluminum to aquatic organisms. *Environ. Toxicol. Chem.* 37 (1): 70–79.

Shen, Q., Truong, L., Simonich, M.T. et al. (2020). Rapid well-plate assays for motor and social behaviors in larval zebrafish. *Behav. Brain Res.* 5 (391): 112625. https://doi.org/10.1016/j.bbr.2020.112625.

United States Environmental Protection Agency (USEPA) (1996, July). *Method 1669 Sampling Ambient Water for Trace Metals at EPA Water Quality Criteria Levels.*

Office of Water, Engineering and Analysis Division. https://www.epa.gov/sites/default/files/2015-10/documents/method_1669_1996.pdf.

United States Environmental Protection Agency (USEPA) (2000, March). *Methods for Measuring the Toxicity and Bioaccumulation of Sediment-Associated Contaminants with Freshwater Invertebrates*. EPA-600-R-99-064, 2e. Washington, DC: Office of Water.

United States Environmental Protection Agency (USEPA) (2001, January). *Water Quality Criterion for the Protection of Human Health: Methylmercury*. EPA-823-R-01-001. Office of Science and Technology, Office of Water. https://www.epa.gov/sites/default/files/2020-01/documents/methylmercury-criterion-2001.pdf.

United States Environmental Protection Agency (USEPA) (2002a, October). *Short-term Methods for Estimating the Chronic Toxicity of Effluents and Receiving Waters to Freshwater Organisms*. EPA-821-R-02-013, 4e. Washington, DC: Office of Water.

United States Environmental Protection Agency (USEPA) (2002b, October). *Methods for Measuring the Acute Toxicity of Effluents and Receiving Waters to Freshwater and Marine Organisms*. EPA-821-R-02-012, 5e. Washington, DC: Office of Water.

United States Environmental Protection Agency (USEPA) (2002c, September). *Guidance for Comparing Background and Chemical Concentrations in Soil for CERCLA Sites*. EPA-540-R-01-003. Washington, DC: Office of Emergency and Remedial Response. https://www.epa.gov/sites/default/files/2015-11/documents/background.pdf.

United States Environmental Protection Agency (USEPA) (2005). *Procedures for the Derivation of Equilibrium Partitioning Sediment Benchmarks (ESBs) for the Protection of Benthic Organisms: Metal Mixtures (Cadmium, Copper, Lead, Nickel, Silver and Zinc) (EPA/600/R-02-011)*. Washington, DC: Office of Research and Development, US Environmental Protection Agency. https://semspub.epa.gov/work/10/500006301.pdf.

United States Environmental Protection Agency (USEPA) (2006, February). Guidance on systematic planning using the data quality objectives process. EPA QA/G-4. EPA/240/B-06/001. https://www.epa.gov/quality/guidance-systematic-planning-using-data-quality-objectives-process-epa-qag-4.

United States Environmental Protection Agency (USEPA) (2007, February). *Aquatic Life Ambient Freshwater Quality Criteria – Copper (2007 Revision)*. EPA 820-R-07-001. Office of Water.

United States Environmental Protection Agency (USEPA) (2008, March). *Procedures for the Derivation of Equilibrium Partitioning Sediment Benchmarks (ESBs) for the Protection of Benthic Organisms: Compendium of Tier 2 Values for Nonionic Organics*. EPA/600/R-02/016. Office of Research and Development. https://semspub.epa.gov/work/10/500006301.pdf.

United States Environmental Protection Agency (USEPA) (2021). *2021 Revision To: Aquatic Life Ambient Water Quality Criterion for Selenium – Freshwater 2016*. EPA 822-R-21-006. Washington, DC: Office of Water. https://www.epa.gov/wqc/aquatic-life-criterion-selenium.

United States Environmental Protection Agency (USEPA) (2024a, December). Water quality standards: establishment of a numeric criterion for selenium for fresh waters in California. EPA 822-R-24-014. https://www.epa.gov/wqs-tech/water-quality-standards-establishment-numeric-criterion-selenium-fresh-waters-california.

United States Environmental Protection Agency (USEPA) (2024b, December). Draft national recommended human health ambient water quality criteria for PFOA, PFOS, and PFBS. https://www.epa.gov/wqc/human-health-water-quality-criteria-pfas.

Villeneuve, D.L., Crump, D., Garcia-Reyero, N. et al. (2014a). Adverse outcome pathway (AOP) development I: strategies and principles. *Toxicol. Sci.* 142 (2): 312–320.

Villeneuve, D.L., Crump, D., Garcia-Reyero, N. et al. (2014b). Adverse outcome pathway (AOP) development II: best practices. *Toxicol. Sci.* 142 (2): 321–330.

Weston, D. and Lydy, M.J. (2012). Stormwater input of pyrethroid insecticides to an urban river. *Environ. Toxicol. Chem.* 31 (7): 1579–1586. https://doi.org/10.1002/etc.

Wild, C.P. (2012). The exposome: from concept to utility. *Int. J. Epidemiol.* 41 (1): 24–32. https://doi.org/10.1093/ije/dyr236.

4

Sequence Effects

Lawrence A. Kapustka

LK Consultancy, Diamond Valley, AB, Canada

The combinations of stressors that organisms are exposed to are infinite. A global inventory prepared in 2020 has more than 350,000 registered chemicals and mixtures identified for commercial use (Wang et al. 2020). Chemical Abstract Services (CAS Content I CAS [accessed 16 April 2023]) has well over 100 million chemicals listed. Any sample of water, sediment, or soil has in addition to some mostly unknowable number of chemicals a myriad of microbes, microscopic plants and animals, and biomolecules including DNA, RNA, and a diverse array of metabolites. Other organisms encountering these mixtures must also contend with numerous physical factors, most notably temperature (Delnat et al. 2019), pH, hardness, dissolved oxygen, and ionizing radiation, to name a few. Initial attempts to understand the consequences of chemical stressors correctly focused on single substances. This despite knowing that the test organisms require dozens of essential nutrients to survive. Standardized protocols were agreed so that otherwise healthy organisms received at least a suite of minimal substances necessary for normal survival and growth. Although much was learned from such experimental tests, including threshold concentrations that result in adverse effects, and dose- or concentration-response relationships, there were situations in which the simple tests were not sufficient. For example, for several metals, response relationships were defined by ratios of coexisting substances rather than the exposure concentration (Alloway 1995, Rietra et al. 2017). Some combinations are protective, whereas others are synergistic. Many other substances, including drugs (see Yin et al. 2014) in combination, display synergistic or antagonistic responses. Typically, substances exhibiting additive effects have a similar mode of action. Synergistic effects can occur if two or more substances target different sites, such as hormone-regulating loci, that then work in combination to yield greater responses.

An even more perplexing experimental challenge has been demonstrated in which the sequence of exposure to stressors alters the response of test organisms (Brooks and Crowe 2019). These experimental findings have implications for predicting responses in real-world settings. And these apply to organism-level dynamics, before even considering the complexity of multispecies interactions in multitrophic systems.

Identifying biological and ecological effects of metals began in earnest in the mid-1800s with work of Liebig (1840) followed by others much later (Horowitz 1945; Epstein 1972; Brooks 1972; McClendon 1976) looking into limiting factors of plant growth. Early in the 20th century, insights into aquatic toxicology of metals were reported (Loeb 1902; Lillie 1904; Mathews 1902, 1903; McGuigan 1903; Woodruff and Bunzel 1909). These were followed by studies of ionoregulation effects during the Dust Bowl era (Smith 1930; Krogh 1937, 1938, 1939). There were important experiments looking into the effects of mixtures of metals and exploration of effects of metal mixtures too (Jones 1939a, b, 1940). The focus before the 1960s was on terrestrial systems and was primarily on nutrient requirements for plants, livestock, and humans. The shift toward the study of toxic effects in aquatic systems was triggered by environmental laws enacted in the United States in the 1970s and beyond. The impressive quantity of data assembled near the end of the 20th century is evidenced by a database for copper with >40,000 reports (ICA 1996).

A confounding aspect in studying toxic effects from mixtures of metal is the essentiality of many metals. Three categories of metals have been identified, namely, those that are essential, those that are beneficial but not essential, and those that are neither essential nor beneficial (Mertz 1981; Alloway 1995). Organizing the vast quantity of data, before the concepts of databases or Big Data were even conceived, was a monumental task. Yet, much of what is known about the effects of metals, including mixtures, was interpreted by Alloway (1974) and Kabata-Pendias and Pendias (1992). Environmental laws triggered a shift in the 1960s and beyond toward investigating nutrient cycling dynamics, pollution concerns, and food-chain transfers of metals from soil, sediment, or water into plants (Woodwell 1967; Howell et al. 1975; Risser et al. 1981; Edmonds 1982; Page et al. 1983; USEPA 1985a, b, c, d, e).

Organisms expend energy to regulate the internal concentrations of these metals (Epstein 1972). Regulation may be internal or occur at the point of uptake across an external membrane. For example, metals may be sequestered into inert granules, bound to organic molecules (e.g., metallothionein in animals or phytochelatins in plants), or shunted into vacuoles away from active enzyme areas. Consequently, in many situations, metals bioaccumulate (USEPA 2002a). Interpretation of toxicity studies on metals that act as essential elements presents a unique challenge because essential elements exhibit classic bell-shaped, or biphasic, dose–response (or exposure–effect) relationships in which adverse effects are observed both at high and low doses, with an optimal mid-range dose (Abernathy et al. 1993). This feature of being essential for growth and having a range of concentrations that is deficient, then sufficient, and then toxic belies a commonly held view in toxicology, which is that the lack of exposure to a substance means that it is not deleterious. Additionally, organisms typically regulate both essential and nonessential metals within their cells, organs, or tissues through metabolic functions. All of these complicate reliance on bioconcentration factors and bioaccumulation factors as a pathway for exposure, primarily because these factors are neither constant nor linear across environmental concentration ranges. This is readily apparent when comparing plant species that have been classified into three categories, namely, accumulators, indicators, or excluders (Baker 1981).

When an organism loses its capacity to regulate the internal concentration of a toxic substance, toxicity occurs. This may manifest in any number of endpoints related in normal growth or survival. Often, this is considered to occur within cells; however, in plants and microorganisms, toxic effects can manifest at the external membrane surface, creating additional challenges in using concentrations in tissues to indicate toxicity.

Another complication in the dogma of toxicity stems from acclimation (organismal) and adaptation (generational) responses (Wallace and Srb 1961). Population of plants growing in mineral-rich environments generally are more tolerant to those substances than are populations of plants growing in other locales (Grime and Hodgson 1969). Bradshaw et al. (1965) reported emergence of metal-tolerant fescue (*Festuca ovina*) and bentgrass within 50–100 years on mine spoils and within 30 years beneath galvanized wire fences. Antonovics et al. (1971) described tolerance in bentgrass (*Agrostis tenuis*) as "evolution in action." Changes of this nature can occur within one generation (Antonovics et al. 1971). They also noted that the tolerant bentgrass was unable to establish in "normal" soils. Similarly, populations of metal-tolerant terrestrial invertebrates have been demonstrated: *Orchesella cincta*, *Porcellio scaber Isotoma notabilis*, and *Onychiurus armatus* (Posthuma and Van Straalen 1993). Intriguing behavioral responses in some wildlife have been correlated with metals in their diets. Selection of nest sites by Abert's squirrels (*Sciurus aberti*) is apparently influenced by the host plant's copper concentration (Snyder and Linhart 1994).

Toxicity tests are generally designed to characterize direct effects at the organismal level (ASTM 1999). In ecological systems, however, indirect effects can play a major role in system dynamics such as community composition, species diversity, nutrient cycling, or rates of litter decomposition and mineralization (Hansen and Tagatz 1980; Meador et al. 1998).

A largely overlooked consideration of toxic effects is that which shapes community composition through food web dynamics. This is despite work done more than eight decades ago! "The world is green" hypothesis wherein top-down control by herbivores ensures that plant growth is kept in check was articulated in 1960 by Hairston et al. This was countered by the bottom-up limitation hypothesis, dubbed "The world is prickly and tastes bad," by Pimm (1991). Pimm argued that plants are adapted to predation, and thus plant availability, energetics, and nutrient limitation ultimately control community composition. Both perspectives were demonstrated experimentally, depending on the target group directly impaired by a toxic substance.

The importance of indirect effects was demonstrated in mesocosm studies by Bothwell et al. (1994) in which they looked at density of diatom communities exposed to solar ultraviolet radiation (SUVR). Direct exposure of diatoms to SUVR reduces photosynthesis and growth of freshwater diatoms. However, there were greater amounts of the algae in exposed complex communities due to greater sensitivity of larval chironomids, consumers of the algae. They argued that predictions of adverse responses to stressors cannot be made from single trophic-level testing.

Although microbes are generally not direct assessment endpoints in risk assessments (Kapustka 1999), they do play significant roles in the condition of plants and animals. For example, mycorrhizal fungi form symbiotic relationships with approximately 80% of dicotyledonous and monocotyledonous plant species and all gymnosperms (Marschner 1995). These associations alter patterns of root growth, influence nutrient and water uptake, protect plants from pathogen, and differentially influence phytotoxic responses (Harley and Smith 1983; Smith and Read 1997). Hyphae can suppress ion uptake from the soil solutions at high concentrations and prevent their translocation into plant tissues. Mycorrhizal fungi increase the number of uptake sites by increasing the effective sorptive surface area of the plant or by altering the affinity constant for a particular metal ion (Cress et al. 1979; Karunaratne et al. 1986; Arnold and Kapustka 1993). In an analogous fashion, gut flora of animals alters the apparent exposure to toxic metals (Todd 1962; Osweiler et al. 1985).

Adaptation of populations to tolerate toxic substances has cascading effects at the community level. An explanatory concept of pollution-induced community tolerance (PICT) was proposed by Blanck et al. (1988). They based it on observations that at contaminated sites sensitive species were replaced by tolerant species. Proceedings of a workshop on PICT stated, "PICT is one of the few tools that can help to demonstrate direct causality between exposure in the environment and effects on in situ microbial communities" (Tlili et al. 2016).

The Community Conditioning Hypothesis (Matthews et al. 1996) posited among other things that history influences the trajectory of ecological systems. Their experimental data demonstrated that the consequences of stressors remained discernable long after previous measures of recovery indicated that the systems had returned to prestress conditions. At issue is the reality that time is not reversible, that systems' trajectories do not and cannot loop back to a previous state. We therefore must adopt new ways of describing management goals pertaining to so-called restoration of affected systems. At best, we can emulate some portion of the goods and services derived from an ecological locale.

In setting targets for "restoration," an understanding of causal chains is crucial for success. Among the earliest observations of the discipline of ecology was Gleason's (1929) recognition that plant communities are shaped by the early colonizers and that chance played a major role in primary and secondary succession. And here, things become exceedingly complex. Research by Jackson et al. (2021) illustrates that the sequence of exposure to different stressors yields different responses at the organism, population, and system level. At the heart of their reasoning is that stressors are not synced to all occur at once. Moreover, the relevant timing or duration of a stressor depends on the life history and life stage of the exposed organisms (Kammenga et al. 2001). They propose that going forward we need to "embrace temporal complexity of multiple stressors and quantify how various realistic asynchronous scenarios will alter their cumulative impacts across different ecosystems."

Each of these factors that complicate interpretation of toxicity test results contribute to the capacity of ecological systems to bounce back from a stress-induced event. Hollings (1973) was the first to characterize the phenomenon in terms of resilience. The basic premise described by Hollings was that systems resided in metaphorical basins and tended to return to a central position within a basin. However, with sufficient stress, the system could be pushed out of a given basin and would undergo change until it relocated into a different metaphorical basin. The description of stability and resilience has been widely accepted within the ecological community. Yet, in general, identifying tipping points that result in a transition to a new basin has been post hoc. A summary article by Capdevila et al. (2021) introduced a special issue in the *Journal of Ecology* that details methods that are based on observation, modelling, or a combination of both that hold promise of being able to predict tipping points, that is, the inflection point at which resilience fails.

Toxicity of metal mixtures differs across a suite of factors including the specific mix of metals, including the ratios of some metals, the relative affinity and potency of the substances vis-à-vis the receptor organisms, and influencing conditions such as pH, reactive carbon, ion strength, and others (Mebane and Balistrieri 2013). Temporal dynamics are critical to the type of response in a complex ecological system due to the ebb and flow and asynchrony of different stressors (Jackson et al. 2021). Jackson et al. (2021) also noted that the sequence of exposure to stressors has implications for the ecological memory embedded in the collective genetic structure of the community. An important variable in the responses to stressors within a community is the array of lifespans within a community.

These interactions documented in experimental and field observations attest to the dynamic nature of social-ecological systems. Clearly, the extensive complexity precludes consideration of the full array of interactions. It would be easy to "toss in the towel" and concede that we can never understand the systems sufficiently to make reasoned environmental management decisions. Yet we are compelled by legal requirements and moral imperatives to provide cost-effective and societally appropriate recommendations to prevent or to mitigate unacceptable risks. Fortunately, over the years, tools to craft environmental management solutions that can gain societal acceptance have been developed. The array of tools includes techniques to gain meaningful input from affected stakeholders (ASTM 2017); guidance on stepwise and iterative development of agreed conceptual models (ASTM 2003); sophisticated analytical methods to acquire data on exposure and effects; computational methods to process complex data to reveal patterns, determine significance, and assign causality (Norton et al. 2009); and transparent ways to communicate risks to diverse stakeholders (Uhlman 2002; Risk Communication Toolkit RCT [itrcweb.org] [accessed 1 February 2024]). That we seldom take advantage of these tools is a serious shortcoming of our practice of environmental risk assessment. Too often, the limited budget is blown on gathering data before knowing what is needed to make the decisions and before stakeholders have been given a chance to express their needs.

Resource managers are faced with impossibly difficult challenges as they attempt to address the issues that residents care about deeply while operating within the constraints of regulatory and policy directives. To illustrate this, examine the skeleton conceptual model in Figure 4.1.

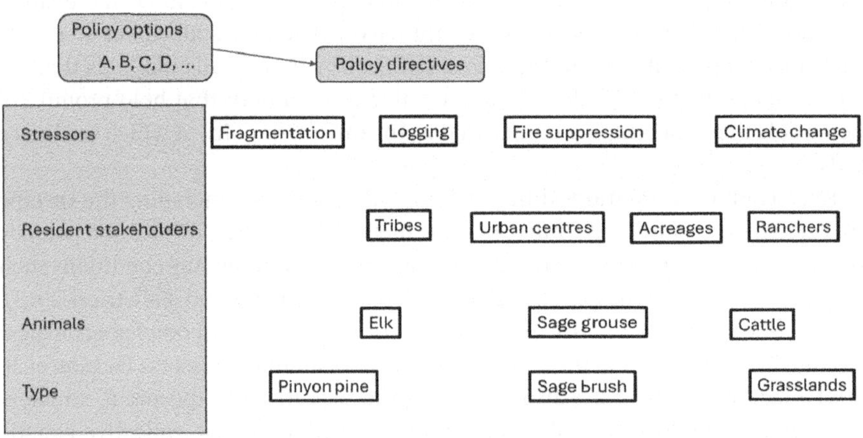

Figure 4.1 Conceptual model depicting the complexity of the social-ecological system influencing Sage Grouse.

- Each stock is on its own trajectory influenced by other entities and stressors.
- Managing for any one stock affects the other entities – some minimally; others dramatically.
- Local and distant stakeholders harbor different values, some incompatible with others.
- Absent active management, trajectories of the various entities will lead to changing configurations over time and space.
- Intensive or intrusive management (e.g., logging) triggers a cascade of changes across the social-ecological landscape.
- Active, adaptive management practices need to be tailored to the dynamics of both the ecological and societal conditions.

Nearly two decades ago, Kapustka (2008) argued that at least 50% of the effort in a risk assessment should be dedicated to the problem formulation segment of the risk assessment paradigm. Of that, the primary focus should be on reaching an agreed conceptual model that incorporates the social-ecological dynamics of the system in question and then specifying the appropriate assessment and measurement endpoints vis-à-vis the management decisions that are to be made. Because affected stakeholders often hold competing values for what they want from the system, achieving an agreed depiction of how the system operates can be and typically is very challenging. Success requires skilled facilitation to reach acceptable trade-offs regarding realistic expectations of the system. It also demands that we avoid imposing our personal values or otherwise skewing the outcome through stealth advocacy or normative science (see Lackey 2001). Finally, as practitioners of risk assessment, we should be humble about what we think we know, in terms of both the ecology of the system and societal desires. Both the ecological and the social considerations are fluid, changing over time. We need to recognize that like the underlying science, risk management options are tentative, and recommendations will require modification as new information or new values emerge. In short, risk assessment should be coupled with adaptive management strategies.

References

Abernathy, C.O., Cantilli, R., and Du, J.T. (1993). Essentiality versus toxicity; some considerations in the risk assessment of essential trace elements. In: *Hazard Assessment of Chemicals* (ed. J. Saxena), 81–113. Washington, DC: Taylor and Francis.

Alloway, B.J. (1974). *Heavy Metals in Soils*. New York: Wiley.

Alloway, B.J. (1995). *Metals in Soils*, 2e. New York: Blackie Academic & Professional, Chapman & Hall.

Antonovics, J., Bradshaw, A.D., and Turner, R.G. (1971). Metal tolerance in plants. *Adv. Ecol. Res.* 7: 1–85.

Arnold, P.T. and Kapustka, L.A. (1993). Comparative uptake kinetics and transport of cadmium and phosphate in *Phleum pratense-Glomus deserticolum* associations. *Environ. Toxicol. Chem.* 12: 177–186.

ASTM (1999). *Standard Guide for Conducting Terrestrial Plant Toxicity Tests.* ASTM Annual Book of Standards. E1963-98, vol. 11.05. Conshohocken: American Society for Testing and Materials.

ASTM (2003). *Standard Guide for Developing Conceptual Site Models for Contaminated Sites.* E1689-95. Conshohocken: American Society for Testing and Materials.

ASTM (2017). *Standard Guide for Framework for a Consensus-Based Environmental Decision-Making Process.* ASTM Annual Book of Standards. E2348-17. Conshohocken: American Society for Testing and Materials.

Baker, A.J.M. (1981). Accumulators and excluders – strategies in the response of plants to metals. *J. Plant Nutr.* 3: 643–654.

Blanck, H., Wangberg, S.A., and Molander, S. (1988). Pollution-induced community tolerance – a new ecotoxicological tool. In: *Functional Testing of Aquatic Biota for Estimating Hazards of Chemicals.* ASTM STP 998 (ed. J. Cairns Jr., and J.R. Pratt), 219–230. Philadelphia: American Society for Testing and Materials.

Bothwell, M.L., Sherbot, D.M.J., and Pollock, C.M. (1994). Ecosystem response to solar ultraviolet-B radiation: influence of trophic level interaction. *Science* 265: 97–100.

Bradshaw, A.D., McNeilly, T.S., and Gregory, R.P.G. (1965). Industrialization, evolution and development of metal tolerant plants. In: *Ecology and the Industrial Society* (ed. G.T. Goodman, R.W. Edwards, and M. Lambert), 327–343. Oxford: Blackwell Press.

Brooks, R.R. (1972). *Geobotany and Biogeochemistry in Mineral Exploration.* New York: Harper & Row.

Brooks, P.R., and Crowe, T.P. (2019). Combined effects of multiple stressors: new insights into the influence of timing and sequence. *Front. Ecol. Evol.* 7:387. https://doi.org/10.3389/fevo.2019.00387.

Capdevila, P., Stott, I., Oliveras Menor, I. et al. (2021). Reconciling resilience across ecological systems, species and subdisciplines. *J. Ecol.* 109: 3102–3113. https://doi.org/10.1111/1365-2745.13775.

CAS Content | CAS (2023) (accessed 16 April 2023).

Cress, W.A., Throneberry, G.O., and Lindsey, D.I. (1979). Kinetics of phosphorus absorption by mycorrhizal and nonmycorrhizal tomato roots. *Plant Physiol.* 64: 484–487.

Delnat, V., Tran, T.T., Verheyen, J. et al. (2019). Temperature variation magnifies chlorpyrifos toxicity differently between larval and adult mosquitoes. *Sci. Total Environ.* 690: 1237–1244.

Edmonds, R.L. (1982). *Analysis of Coniferous Forest Ecosystems in the Western United States*, US/IBP Synthesis Series #14. Stroudsburg: Hutchinson Ross Publishing Company.

Epstein, E. (1972). *Mineral Nutrition of Plants: Principles and Perspectives*. New York: Wiley.

Gleason, H.A. (1929). The significance of Raunkiaer's law of frequency. *Ecology* 10: 406–408. https://doi.org/10.2307/1931149.

Grime, J.P. and Hodgson, J.G. (1969). An investigation of the ecological significance of lime-chlorosis by the means of large-scale comparative experiments. In: *Ecological Aspects of the Mineral Nutrition of Plants* (ed. I.H. Rorison), 67–99. Oxford: Blackwell Press.

Hansen, D.J. and Tagatz, M. (1980). A laboratory test for assessing impacts of substances on developing communities of benthic estuarine organisms. *ASTM Spec. Tech. Publ.* 27408S: 40–57. https://doi.org/10.1520/STP27408S.

Harley, J.L. and Smith, S.E. (1983). *Mycorrhizal Symbiosis*. New York: Academic Press.

Hollings, C.S. (1973). Resilience and stability of ecological systems. *Annu. Rev. Ecol. Syst.* 4: 1–23.

Horowitz, N.H. (1945). On the evolution of biochemical syntheses. *Proc. Natl. Acad. Sci.* 31: 153–157.

Howell, F.G., Gentry, J.B., and Smith, M.H. (1975). *Mineral Cycling in Southeastern Ecosystems*. Springfield: U.S. Energy Research and Development Administration, NTIS.

ICA (International Copper Association) (1996). The biological importance of copper: Final report. ICA Project No. 223. New York.

Jackson, M.C., Pawar, S., and Woodward, G. (2021). The temporal dynamics of multiple stressor effects: from individuals to ecosystems. *Trends Ecol. Evol.* 36: 402–410. https://doi.org/10.1016/j.tree.2021.01.005.

Jones, J.R.E. (1939a). Antagonism between salts of the heavy and alkaline-earth metals in their toxic action on the tadpole of the toad, *Bufo bufo* (L). *J. Exp. Biol.* 16: 313–333.

Jones, J.R.E. (1939b). The relation between the electrolytic solution pressures of the metals and their toxicity to the stickelback (*Gasterosteus aculeatus* L.). *J. Exp. Biol.* 16: 425–437.

Jones, J.R.E. (1940). A further study of the relation between toxicity and solution pressure with *Polycelis nigra* as test animal. *J. Exp. Biol.* 17: 408–415.

Kabata-Pendias, A. and Pendias, H. (1992). *Trace Elements in Soils and Plants*, 2e. Boca Raton: CRC Press.

Kammenga, J.E., Van Gestel, C.A.M., and Hornung, E. (2001). Switching life-history sensitivities to stress in soil. *Ecol. Appl.* 11 (1): 226–238.

Kapustka, L.A. (1999). Microbial endpoints: The rationale for their exclusion as ecological assessment endpoints. *Hum. Ecol. Risk. Assess.* 5: 691–696.

Kapustka, L. (2008). Limitations of the current practices used to perform ecological risk assessment. *Integr. Environ. Assess. Manag.* 4: 290–298.

Karunaratne, R.S., Baker, J.H., and Barker, A.V. (1986). Phosphorus uptake by mycorrhizal and nonmycorrhizal roots of soybean *Glycine max*. *J. Plant Nutr.* 9: 103–1314.

Krogh, A. (1937). Osmotic regulation in the frog *(Rana esculenta)* by active absorption of chloride ions. *Scand. Arch. Physiol.* 76: 60–73.

Krogh, A. (1938). The active transport of ions in some freshwater animals. *Z. Vergleich. Physiol.* 25: 335–350.

Krogh, A. (1939). Osmotic regulation in aquatic animals [originally published by Cambridge University Press and reprinted in English in 1965 in an unabridged and unaltered edition by Dover Publications, Inc., New York, 242 pp].

Lackey, R.T. (2001). Values, policy, and ecosystem health. *Bioscience* 51: 437–443.

Liebig, J. (1840). *Chemistry in Its Agriculture and Physiology*. London: Taylor and Walton.

Lillie, R.S. (1904). The relation of ions to ciliary movement. *Am. J. Phys.* 10: 419–443.

Loeb, J. (1902). Studies on the physiological effects of the valency and possibly the electrical charges of ions. I. The toxic and anti-toxic effects of ions as a function of their valency and possibly their electrical charge. *Am. J. Phys.* 6: 411–433.

Marschner, H. (1995). *Mineral Nutrition of Higher Plants*, 2e. San Diego: Academic Press.

Mathews, A.P. (1902). The nature of nerve stimulation and of changes in irritability. *Science* 15: 492–498.

Mathews, A.P. (1903). The nature of nerve irritability and of chemical and electrical stimulation. Part II. *Science* 17: 729–733.

Matthews, R.A., Landis, W.G., and Matthews, G.B. (1996). The community conditioning hypothesis and its application to environmental toxicology. *Environ. Toxicol. Chem.* 15: 597–603. https://doi.org/10.1002/etc.5620150427.

McClendon, J.H. (1976). Elemental abundance as a factor in the origins of mineral nutrient requirements. *J. Mol. Evol.* 8: 175–195.

McGuigan, H. (1903). The relation between the decomposition tension of salts and their antifermentative properties. *Am. J. Phys.* 25: 444–451.

Meador, J.P., Sible, T.H., Swartzman, G.L., and Taub, F.B. (1998). Copper tolerance by the freshwater algal species *Oocystis pusilla* and its ability to alter free-ion copper. *Aquat. Toxicol.* 44: 69–82.

Mebane, C.A. and Balistrieri, L.S. (2013). Predicting the toxicity of metal mixtures. *Sci. Total Environ.* 466-437C: 788–799. https://doi.org/10.1016/j.scitotenv.2013.07.034.

Mertz, W. (1981). The essential elements. *Science* 213: 1332–1338.

Norton, S.B., Cormier, S., Suter, G.W. et al. (2009). Causal analysis/diagnosis decision information system (CADDIS). In: *Decision Support Systems for Risk-Based Management of Contaminated Sites* (ed. A. Marcomini, G.W. Suter, and A. Critto),

351–374. Springer Science Business Media, LLC https://doi.org/10.1007/978-0-387-09722-0.

Osweiler, G.D., Carson, T.L., Buck, W.B., and Van Gelder, G.A. (1985). Selenium. In: *Clinical and Diagnostic Veterinary Toxicology* (ed. G.D. Osweiler, T.L. Carson, W.B. Buck, and G.A. Van Gelder), 132–142. Dubuque: Kendall/Hunt Publishing Co.

Page, A.L., Gleason, T.L., Smith, J.E. et al. (1983). Utilization of municipal wastewater and sludge on land. Proceedings of the 1983 Workshop. University of California, Riverside, CA.

Pimm, S.L. (1991). *The Balance of Nature: Ecological Issues in the Conservation of Species and Communities*. Chicago and London: University of Chicago Press.

Posthuma, L. and Van Straalen, N.M. (1993). Heavy-metal adaptation in terrestrial invertebrates: a review of occurrence, genetics, physiology and ecological consequences. *Comp. Biochem. Physiol. C* 106: 11–38.

Rietra, R.P.J.J., Heinen, M., and Dimkpa, C.O. et al. (2017). Effects of nutrient antagonism and synergism on yield and fertilizer use efficiency. *Commun. Soil Sci. Plant Anal.* 48:1895–1920, https://doi.org/10.1080/00103624.2017.1407429.

Risk Communication Toolkit RCT (2024). itrcweb.org (accessed 1 February 2024).

Risser, P.G., Birney, E.C., Blocker, H.D. et al. (1981). *The True Prairie Ecosystem*, US/IBP Synthesis Series #16. Stroudsburg: Hutchinson Ross Publishing Co.

Smith, H.W. (1930). The absorption and excretion of water and salts by marine teleosts. *Am. J. Phys.* 93: 480–505.

Smith, S.E. and Read, D.J. (1997). *Mycorrhizal Symbiosis*, 2e. San Diego: Academic Press.

Snyder, M.A. and Linhart, Y.B. (1994). Nest-site selection by Abert's squirrel: Chemical characteristics of nest trees. *J. Mammal.* 75 (1): 136–141.

Tlili, A., Berard, A., Blanck, H. et al. (2016). Pollution-induced community tolerance (PICT): towards an ecologically relevant risk assessment of chemicals in aquatic systems. *Freshw. Biol.* 61: 2141–2151. https://doi.org/10.1111/fwb.12558.

Todd, J.R. (1962). Chronic copper poisoning in farm animals. *Vet. Bull.* 32: 573–580.

Uhlman, K. (2002). Environmental risk communication. Environmental Protection (Environmental Risk Communication – Environmental Protection [eponline.com] [accessed 1 February 2024]).

USEPA (1985a). *Guidelines for Deriving Numerical National Water Quality Criteria for the Protection of Aquatic Organisms and Their Uses*. EPA/822/R-85/100, 98. Duluth: Office of Research and Development.

USEPA (1985b). Ambient water quality criteria for lead. USEPA 440/5-84-027, Washington, DC.

USEPA (1985c). *Environmental Profiles and Hazard Indices for Constituents of Municipal Sludge: Cadmium*. Washington, DC: Office of Water Regulations and Standards.

USEPA (1985d). *Environmental Profiles and Hazard Indices for Constituents of Municipal Sludge: Copper*. Washington, DC: Office of Water Regulations and Standards.

USEPA (1985e). *Environmental Profiles and Hazard Indices for Constituents of Municipal Sludge: Lead*. Washington, DC: Office of Water Regulations and Standards.

USEPA (2002a). Development of a framework for metals assessment and guidance for characterizing and ranking metals; draft action plan. EPA/630/P-02/003A, Washington, DC.

Wallace, B. and Srb, A. (1961). *Adaptation*. Edgewood Cliffs: Prentice Hall.

Wang, Z., Walker, G.W., Muir, D.C.G., and Yoshida, K.N. (2020). Toward a global understanding of chemical pollution: a first comprehensive analysis of national and regional chemical inventories. *Environ. Sci. Technol.* 54: 2575–2584. https://doi.org/10.1021/acs.est.9b06379.

Woodruff, L.L. and Bunzel, H.H. (1909). The relative toxicity of various salts and acids toward paramecium. *Am. J. Phys.* 25: 190–194.

Woodwell, G.M. (1967). Toxic substances and ecological cycles. *Sci. Am.* 216: 24.

Yin, N., Ma, W., Pei, J. et al. (2014). synergistic and antagonistic drug combinations depend on network topology. *PLoS One* 9 (4): e93960. https://doi.org/10.1371/journal.pone.0093960.

5

Addressing the Challenges of Pharmaceutical Pollution

Samuel A. Welch

Norwegian Institute for Water Research (NIVA), Ecotoxicology and Risk Assessment, Oslo, Norway

Introduction

In their 2009 and subsequent 2015 paper, the Stockholm Resilience Centre (SRC) identified nine planetary boundaries – limits of the earth's ability to accommodate human environmental impacts (Rockström et al. 2009; Steffen et al. 2015). More recently, a 2022 reassessment suggested humanity is transgressing safe operating spaces in six of these areas: biogeochemical flows (nitrogen and phosphorus pollution), land-system change, freshwater change, genetic biodiversity of the biosphere, climate change, and novel entities – that is to say, all other pollutants (Persson et al. 2022; Wang-Erlandsson et al. 2022).

The simultaneous transgression of these boundaries indicates a complex stressor landscape for humans and the environment. Society and ecosystems must adapt to a rapidly changing climate, landscape, and biosphere: a problem space where problems and solutions are intimately linked, poorly understood, complex, and dynamic (Allen 2013) – archetypal *wicked problems* (Lertxundi et al. 2020). Simple solutions to this interlocking set of wicked problems are out of the question, and any effective approach to environmental challenges must be able to account for uncertainty, the future, and the interactions between issues often considered in separate, siloed fields.

The pharmaceutical pollution of water sits at the intersection of several planetary boundaries. On one side, the WHO has named climate change as the "single biggest health threat facing humanity" (WHO 2021), while on the other side, healthy biodiversity and ecosystems are also important contributors to human health (Myers and Patz 2009; WHO and CBD 2015). Pharmaceuticals are a beneficial and crucial tool in the management of human health risks. As climate change (WHO 2021), pollution (Fuller et al. 2022), environmental degradation (WHO and CBD 2015; Chiabai et al. 2018), and population change (van der Aa and Kommer 2010) drive growing health needs, the environmental impacts of pharmaceutical consumption will grow.

Comparative rankings of the most prominent global risks to humanity today tend to focus on broader scale issues such as climate change and extreme weather, biodiversity loss, and large-scale environmental damage incidents (World Economic Forum 2023). Pharmaceuticals do not fit neatly into this framework, but their presence in the environment has driven notable local population collapses – e.g., vultures and diclofenac (Swan et al. 2006) and developmental disruption (Jobling et al. 1998). Pharmaceuticals occupy a particularly interesting role in patterns of environmental impact and organism health. Where environmental impact drives degradation to health-relevant ecosystem services (Gerbersdorf et al. 2011; Topaz et al. 2020) (including the disposal of cattle carcasses by the aforementioned vultures), demand for pharmaceuticals may grow, risking the creation of deleterious feedback effects.

Risk mitigation and management decisions in such a complex, dynamic system require an assessment of how environmental risks may change in the future. In environmental protection, as well as in health, *prevention is better than cure* (Inger 2021). In order to manage global health and environmental quality in as equitable, cost-effective, and scientifically informed manner as possible, it is necessary to foresee future risks, rather than simply assess present impacts (Boxall et al. 2012).

In this chapter, we discuss drivers of present and future pharmaceutical risk in the environment. We summarize existing approaches to the key challenges of assessing environmental risk of pharmaceuticals and how we developed an object-oriented Bayesian Network (OOBN) prototype in order to address some of the challenges identified. Finally, we discuss future possibilities for extending our prototype to address other potentially relevant aspects of global change.

Drivers of Pharmaceutical Risk

Pharmaceutical pollution arises in the environment via several linked pathways (Figure 5.1). Economically, the demand for pharmaceuticals drives their manufacture (Figure 5.1a). The manufacture of pharmaceuticals may result in their emission to the environment (Figure 5.1b) (Bielen et al. 2017), depending on the level of pollution control at manufacturing sites. A fraction of the pharmaceuticals consumed by human and veterinary end users is excreted to the environment as both unmodified active pharmaceutical ingredient (API) and transformation products entering the environment either directly or at some stage along formal or informal wastewater treatment infrastructure. Finally, pharmaceuticals that are not consumed may enter the environment via disposal down the drain, leaching from landfills, and other disposal pathways.

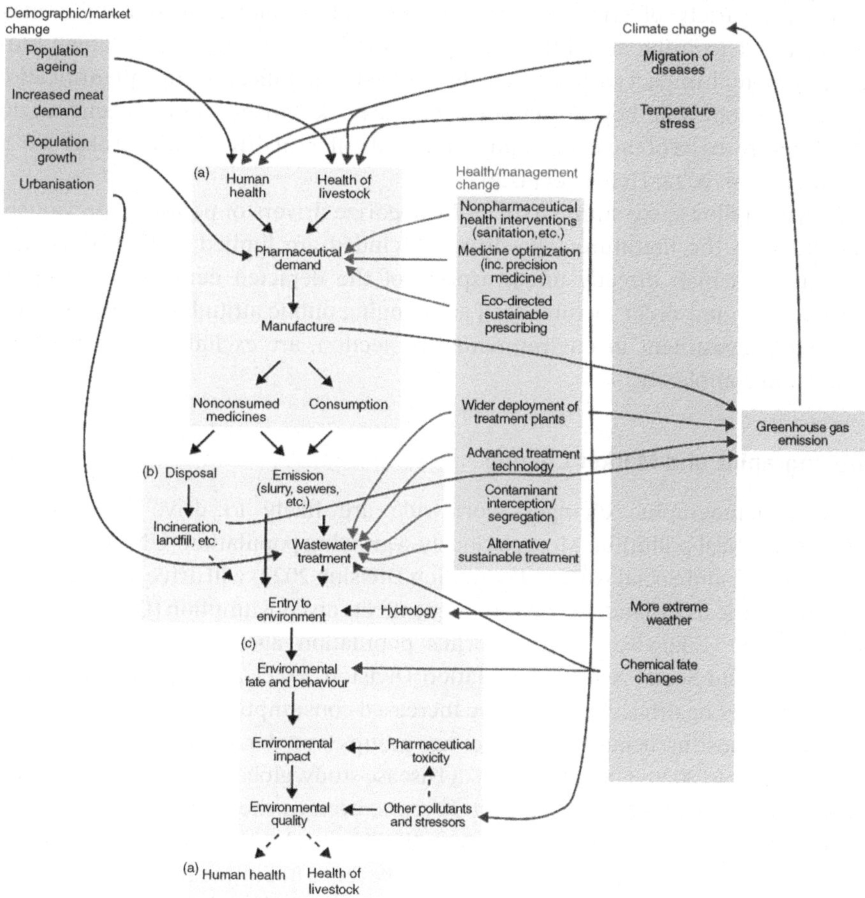

Figure 5.1 A conceptual, nonexhaustive diagram of the first-order influence of health issues on pharmaceutical consumption (a), routes of pharmaceutical emission to the environment (b), and environmental impacts (c); and potential future drivers of change: demographic/market, health/management, and climate change. Original diagram, based on. *Source:* Boxall et al. (2012), Bunke et al. (2019), and Helwig et al. (2023).

Once pharmaceuticals and their degradation products are in the environment, their fate depends on both the environment and their intrinsic physicochemical properties (Figure 5.1c) (Bavumiragira et al. 2022). These chemicals may subsequently cause environmental impacts by toxicity to wildlife, individually, or, more realistically, in interaction with other pollutants (both pharmaceutical and nonpharmaceutical) and stressors (Pirotta et al. 2022). Eventually, changes to environmental quality may then complete the feedback loop by driving changes to the health of humans and livestock (Figure 5.1a).

A great variety of environmental, social, and technological dynamics may be expected to influence different aspects of this pharmaceutical demand-to-environmental-impact pathway. A truly exhaustive qualification and quantification of these factors is far beyond the scope of this book. Furthermore, in general, no clear picture exists of the relative importance of different drivers (Boxall et al. 2012; Sumpter et al. 2022; Helwig et al. 2023).

In what follows, we summarize and categorize drivers of pollution previously identified in the literature. The drivers included are limited to first-order factors, those which directly affect aspects of the depicted demand-to-pollution pathway. Second-order factors, such as changing public attitudes driving greater, or lesser, investment in environmental protection, are excluded for reasons of space and complexity.

Demographic and Market Change

Several demographic changes worldwide are likely to drive increases in pharmaceutical pollution. Most obviously, a growing population (UN Department of Economic and Social Affairs, Population Division 2022) will drive an increase in pharmaceutical demand, with consequent effects on consumption (González Peña et al. 2021). Likewise, as the world's population ages (UN Department of Economic and Social Affairs, Population Division 2022), an increased incidence of age-related health issues will drive increased consumption of pharmaceuticals, especially those associated with older age groups (van der Aa et al. 2011).

According to 2020s Global Burden of Disease study, global overall human health has steadily improved over the past 30 years, even as the burden of disease has shifted to conditions associated with later life (Vos et al. 2020). The effect this trend will have on overall pharmaceutical use, however, is not immediately apparent, as different diseases and conditions will be treated with different types and quantities of pharmaceuticals.

A growing population, in association with economic development, will also create greater demand for food, including livestock, especially in newly developing nations. This in turn will drive increased veterinary pharmaceutical use (Van Boeckel et al. 2015, 2019). Urbanization may also be driven by, or be a driver of, economic development (Shaban et al. 2022). Urbanization may drive pollution (Liang and Yang 2019), positively or negatively affect the health of urban citizens (Godfrey and Julien 2005), overwhelm existing wastewater infrastructure (Osuhor and Essien 1978), and/or improve access to wastewater treatment by reducing the relative per-capita cost of constructing infrastructure (e.g., sewage lines) (Massoud et al. 2009).

Health and Management Change

The medical needs of populations will be a clear and obvious driver of pharmaceutical consumption, and thus emission to the environment, in the future. For the sake of focus, however, we will restrict our attention to environmentally relevant health and management changes.

Helwig et al. (2023) identify a number of ongoing trends in treatment development, improvement, and deployment that could, for an equal disease burden, reduce pharmaceutical consumption. Nonpharmaceutical interventions (health promotion, green prescribing, etc.) can reduce both communicable and noncommunicable disease burden via a wide variety of approaches and displace existing pharmaceuticals (WHO 2022). Prescriptions can be better optimized to individual patient needs (NICE Medicines and Prescribing Centre [UK] 2015) and environmental hazard reduction (Daughton 2014). Likewise, disposal can be improved through greater public awareness and compliance with good practice (Daughton 2014; Vatovec et al. 2017) and better infrastructure at the point of disposal (incinerators, landfills) (Cook et al. 2012), reducing emission to the environment. Pharmaceuticals could even be "recycled" at the pharmacy or institutional level to reduce wastage (McRae et al. 2016).

An immediately intuitive method of removing pharmaceuticals from waste streams is interception and removal via wastewater treatment technology, the so-called "end of pipe" treatment. Wastewater treatment plants (WWTPs) use a variety of technologies to removal various pollutants, including organic micropollutants such as pharmaceuticals. The broad diversity of pharmaceutical pollutants means that removal rates can vary considerably (Courtier et al. 2019), so multiple technologies may be required for effective risk mitigation. To their advantage, implementing these technologies may have the additional benefit of reducing risk from other pollutants; however, a number of potential disadvantages remain. Technological solutions can present high development and operational costs (Logar et al. 2014) and require considerable energy to use (Longo et al. 2016; Wang et al. 2016). Furthermore, removal of pollutants at WWTPs does not necessarily imply a reduction in environmental risk. In some cases, pollutants may be transformed to *more hazardous* substances (Li et al. 2016), or harmful by-products released (Ikehata 2019). Alternatively, where harmful substances are removed to be partitioned to sludge, they may still enter the environment by the later reuse of said sludge – for example, as agricultural fertilizer (Bolesta et al. 2022). Alternative, more sustainable wastewater treatment technologies, such as constructed wetlands, have been successfully used to treat water contaminated with pollutants, but their adoption has remained on a far smaller scale than traditional infrastructure (Li et al. 2014; Verlicchi and Zambello 2014).

Climate Change

Anthropogenic climate change is not the only global-scale environmental risk. However, it is generally identified by relevant groups and authorities as the most serious medium-to-long-term threat to economies (World Economic Forum 2023), human health (WHO 2021), and a serious threat to wildlife (Bellard et al. 2022).

Climate change has the potential to interact with the demand-to-impact pipeline at all levels. Human and livestock health can be negatively affected by growing heat stress (Lundgren et al. 2013; Bagath et al. 2019), by changing patterns of exposure to diseases (Mora et al. 2022), by droughts and attendant water stress (Cook et al. 2018), and by the impacts of extreme weather events (de' Donato and Michelozzi 2014).

More extreme weather may also affect pollution impact via changing hydrological patterns in aquatic ecosystems. As climate change alters the size and flow rate of receiving bodies of water, concentrations of pharmaceuticals in the environment will change due to effects of dilution (Muñoz et al. 2009; Bunke et al. 2019). Depending on the local changes to weather patterns, average dilution may increase or decrease, with corresponding effects on concentrations. However, where weather becomes more extreme and unpredictable, so too will dilution, making for more uncertain concentrations, and higher acute exposure (Osorio et al. 2014). Climate change is also expected to affect the global distribution, transport, and fate of pollutants in the environment (Ma et al. 2016), particularly persistence (Kong et al. 2013).

Modeling the Risks of Pollutant Mixtures

Climate change–driven stressors are likely to also affect wildlife and ecosystems (Malhi et al. 2020) in complex and difficult-to-predict ways (Ormerod et al. 2010; Jackson et al. 2016). The issue of multiple stressor interactions, and the subissue of the toxicity of mixtures of chemicals, has long been a bugbear of the ecotoxicology world due to the difficulties of translating complex, multistressor environments into systems that can be practically studied under controlled settings (Pirotta et al. 2022). Mixture effects between stressors are typically greater than the effects of each individual component, and thus even mixtures of pharmaceuticals (and other pollutants) at low concentrations can pose considerable ecotoxicity (Backhaus 2014). A number of mixture risk approaches have been developed that aim to be simple enough to be practical, but complex enough to be protective (Pirotta et al. 2022), but no one method has yet obtained broad acceptance.

Risk is often quantified in environmental risk assessment (ERA) by the calculation of a unitless risk score, typically including information about the magnitude and probability of a given risk. The most common such score is Risk Quotient (RQ), or Risk Characterization Ratio. RQ is calculated by first describing the predicted toxicity of a substance, in the form of a Predicted No Effect Concentration (PNEC), the concentration below which it has no environmental effect. This PNEC is then used to divide a Predicted or Measured Environmental Concentration (PEC or MEC), to give a final RQ value. RQs are intended for the comparison of relative risks of, for instance, similar chemicals, allowing for substitution to reduce environmental impact (Leeuwen 2007). RQs are extremely easy to calculate and, especially in the EU, have become a core component of the ERA paradigm (De Bruijn et al. 2002). Consequently, a large proportion of the publicly available information on the environmental risk of chemicals, including pharmaceuticals, is available as RQs.

Despite their acceptance in Europe, however, RQs have received considerable criticism. Jager et al. (2001) and more recently Raimondo and Forbes (2022), point out that RQs provide no clear information on the probability of a given adverse outcome occurring, or of the outcome's magnitude. Landis and Chapman (2011) criticize no observed effect levels (and by extension PNECs and RQs) for failing to capture the fundamental nonlinearity of the dose–response relationship between stressors and their effects, while Hope (2009) criticizes the use of "bright line" regulatory thresholds in general. Furthermore, the calculation of PNECs involves the selection of an assessment factor designed to account for uncertainty by penalizing (i.e., increasing) the toxicity of a substance with less study data available (De Bruijn et al. 2002). Although this may provide an effective incentive for stakeholders to fund studies to reduce predicted risk, it creates a situation where risk management decisions, uncertainty (Moe et al. 2022), probability, and magnitude (Fairbrother et al. 2016) are baked into a single score that is often accepted as a *fait accompli* not only for screening-level assessments (Yu et al. 2017), its original purpose, but also as a general, mathematically independent quantification of risk.

Despite these failings, RQs have also seen development into a score of the summed risk of multiple substances. Sum of Risk Quotients (ΣRQ) is subject to all the criticisms of individual RQs and more, and the summing of ratios (with different numerators and denominators) is to say the least controversial (Mochon 1993). ΣRQ has nevertheless seen uptake as a simple-to-calculate value for comparison between scenarios and, perhaps most notably, has been further developed into the Mixture Assessment Factor (MAF), developed by Backhaus for the Swedish Chemicals Agency (2021). An in-depth explanation is left to the original author, but in summary: For a given mixture of n chemicals, a ΣRQ is calculated assuming that all individual components are mitigated to $RQ \leq 1$. If the subsequent ΣRQ of this mixture still exceeds 1, then an MAF (typically around $n/2$) is applied stepwise to components until the ΣRQ is below 1.

Modeling the Risks of Multiple Stressors

Although the prediction of mixture risks to ecosystems remains a difficult issue, integrating changes in other, nonchemical stressors and risk drivers is more challenging. Where factors are anticipated to directly drive changes in pharmaceutical consumption (e.g., population change), these can be easily incorporated into ERA. However, phenomena such as the interactions between temperature changes and pollutant effects (Mangold-Döring et al. 2022) are harder to integrate into existing paradigms.

Forecasting of the conditions that drive future environmental risk is typically presented in the form of sets of scenarios. Scenarios group possible outcomes of various factors into sets of plausible, thematic directions. To date, extensive work with the development of plausible scenarios of future change in the field of climate change research has shown their value as risk assessment and communication tools (van Vuuren et al. 2012; O'Neill et al. 2014). Various scenarios relevant to pharmaceutical pollution have been developed, including coverage of climate change (IPCC 2022), socioeconomic development (O'Neill et al. 2014), chemical emissions (Desrousseaux et al. 2022), and demographic development (Thomas and Tømmerås 2022). However, the integration of scenarios with existing ERAs remains largely an academic practice and has yet to be incorporated into regulatory requirements.

Bayesian Networks for Environmental Risk Assessment

Bayesian networks (BNs) (Figure 5.1a,b) are directed acyclic graphs (DAGs) where probability-distributed variables are depicted as nodes and directional relationships as edges. The relationships between nodes are described by Conditional Probability Tables (CPTs) that set the probability of each state of a child node depending on the state of a parent node. OOBNs (Figure 5.1c) are a subclass of BNs that use an object-oriented programming paradigm to represent repeated motifs, such as cyclic processes or common pathways (Koller and Pfeffer 1997) (Figure 5.2).

BNs have been applied to ERA for the past 20 years, with proponents citing their ability to explicitly quantify uncertainty and variability (Hart and Pollino 2008). BNs have seen some use in regulatory settings, particularly for ecosystem management (Mitchell et al. 2021; Adams et al. 2022), but have yet to be broadly used by stakeholders. Likewise, the use of OOBNs for addressing environmental issues has thus far been extremely limited (Carmona et al. 2011; Lehikoinen 2014).

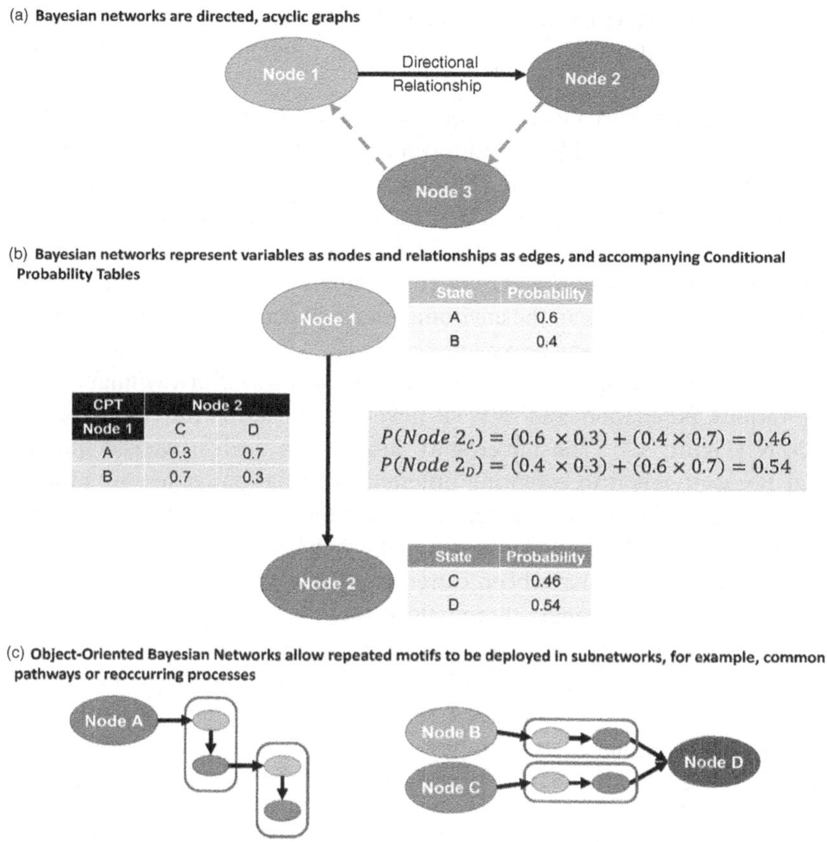

Figure 5.2 (a) Bayesian networks are directed acyclic graphs (DAGs), i.e., the relationships between nodes have direction, and nodes and edges cannot form a loop. (b) Variables are represented as nodes with different possible states (or a continuous distribution of state) that have different probabilities. In the example, the state of Node 2 depends on the state of Node 1. The specific probabilities of these are depicted in a Conditional Probability Table (CPT). When a Bayesian network is compiled, the probability of each state of every node is calculated iteratively using these CPTs and the states of prior nodes. (c) Object-oriented Bayesian networks allow repeated motifs of nodes (represented by rounded rectangles) to be deployed in subnetworks.

BNs for ERA: A Norwegian Example

Norway, a highly developed nation in the north of Europe, faces relatively low pharmaceutical pollution risk (Bouzas-Monroy et al. 2022; Wilkinson et al. 2022).

However, its extensive pharmaceutical sales data collection (Sommerschild et al. 2021) makes it a useful case study for the prediction of pharmaceutical pollution.

Additionally, as a part of the European Free Trade Association, Norway is likely to be heavily affected by the European Union (EU)'s proposed new Urban Wastewater Treatment Directive (European Commission 2022). This would require considerable upgrades to WWTPs above a certain size, which has been criticized by industry groups (EurEau 2023; Norsk Vann 2023) as expensive and potentially with adverse environmental impacts of its own. This raises an interesting question: how can the environmental and economic costs and benefits of such legislation in Norway be weighed?

In order to begin to address how pharmaceutical risk in Norway might evolve under various population and WWTP development scenarios, we developed an OOBN (Figure 5.3). The uncertainty intrinsic in such a wicked question made a BN well suited to exploring different potential outcomes. In addition, by using an object-oriented approach, we were able to account for considerable variation between the toxicity and consumed weight of different pharmaceutical pollutants, which would have been extremely discretized into environmentally relevant intervals using a single common pathway.

Using this OOBN, we predicted the surface water risks of a panel of 6 high-risk pharmaceuticals in Norway. We combined three population growth scenarios, three WWTP upgrade scenarios, two timepoints (2020 and 2050), and three Norwegian counties picked as examples of rural, semiurban, and urban areas. Notwithstanding criticisms earlier in this chapter, RQs were calculated per API and scenario as the only risk metric applicable with available toxicity data for the APIs. To allow for broad comparisons of different scenarios, the distribution of ΣRQs was calculated. Finally, a joint probability of any of the APIs exceeding an arbitrary RQ threshold (set to 100 and 500 in this example) was calculated.

Our results, summarized here, show the effects of different scenarios on the sum of RQs. Using our OOBN, we were able to quickly calculate probability distributions under different population (Figure 5.5) and wastewater treatment scenarios (Figure 5.4). Under fixed WWT conditions, predicted overall risk (ΣRQ) increased nonuniformly with population growth. In an urban county, the probability of high ΣRQ increased in 2050, but remained uniformly high across all three population scenarios. The rural county showed the most pronounced increase in the probability of high RQs, with $P(\Sigma RQ > 1{,}000)$ exceeding 75% under the highest population growth scenarios. The semiurban county also showed a considerable increase in the probability of higher summed risk, although the discretization of the network was not sufficiently sensitive to detect changes between main and high population growth.

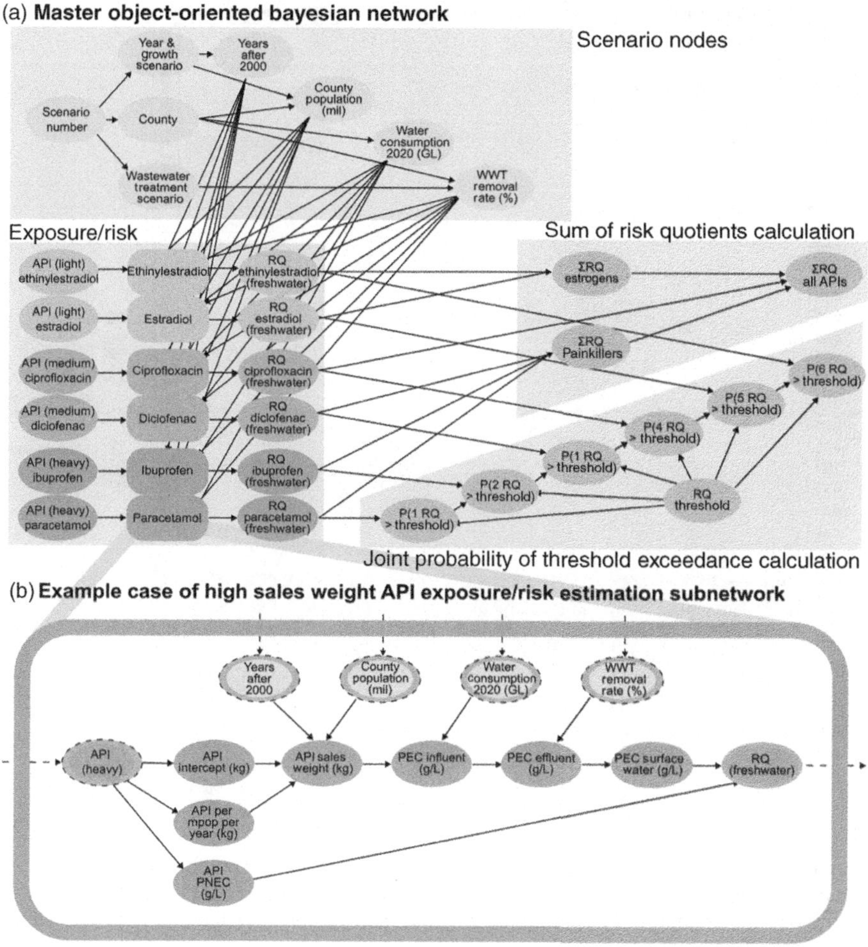

Figure 5.3 Condensed diagram of an object-oriented Bayesian network used for the prediction of pharmaceutical environmental risk. The main network (a) consists of a number of scenario nodes, six API (active pharmaceutical ingredient) exposure and risk calculation subnetworks corresponding to low, medium, and high sales weight APIs, and 10 risk nodes (orange) that calculate both sums of Risk Quotients (RQs) and probability of exceeding various RQ thresholds, respectively, by category and by number of APIs. Rounded rectangles indicate instances of subnetworks, containing API information and a pipeline for forecasting sales weights and predicting environmental concentrations and subsequent risk. Whole-line ovals represent ordinary nodes. Dotted ovals with gray insets represent input nodes, which receive inputs in subnetwork instances from regular nodes. Whole-line ovals with gray insets represent output nodes, which provide outputs from subnetworks to the main network. *Source:* Welch et al. (2023) / OSF Preprints / CC BY 4.0.

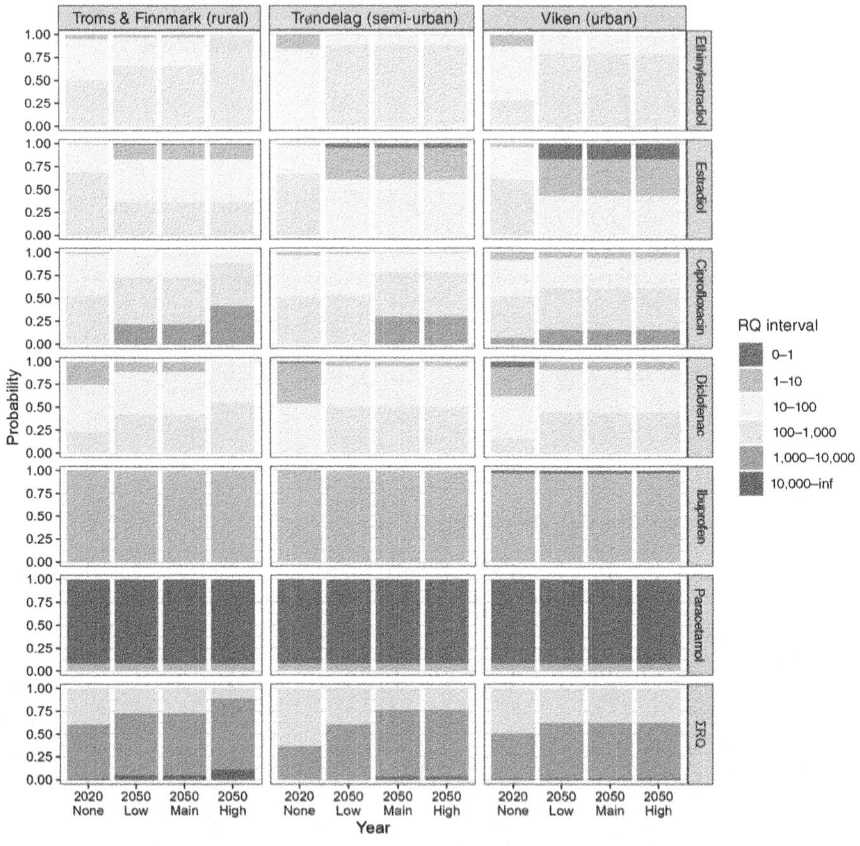

Figure 5.4 Predicted distributions of Risk Quotients (RQs) for six individual pharmaceuticals and all pharmaceuticals summed across three counties and four combinations of year and population growth scenario. *Source:* Welch et al. (2023) / OSF Preprints / CC BY 4.0.

Under fixed population conditions (Figure 5.4), risk distributions under the current WWTP scenario showed the lowest probability of high ΣRQ values in the urban county, due to existing advanced treatment technology. However, under upgrade scenarios a pronounced reduction in risk distributions was seen in the rural county, where high removal rates and dilution in wastewater combined to limit the risk distributions of APIs.

Overall risk (ΣRQ) was notably driven by a subset of APIs (ethinylestradiol, diclofenac, and especially the antibiotic ciprofloxacin) under all scenarios. By contrast, the painkillers ibuprofen and paracetamol showed relatively low risk ($RQ < 10$) under all scenarios. ΣRQs were also compared to the joint probability that any API exceeded a set RQ threshold (of 100 or 500). Both indicators showed

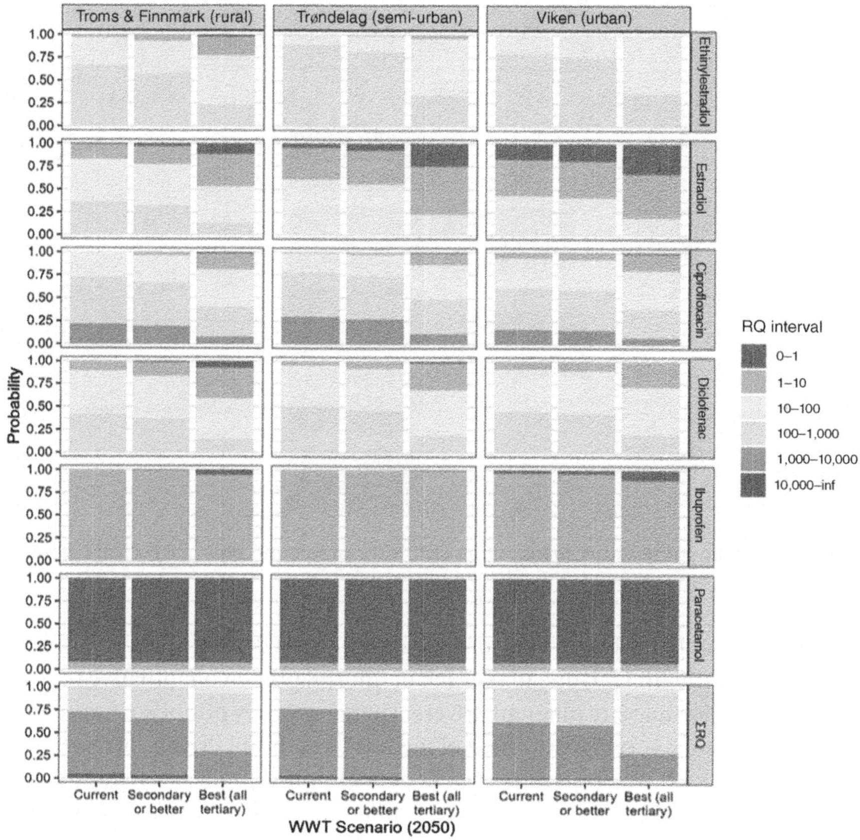

Figure 5.5 Comparison of Risk Quotient (RQ) distributions for six individual pharmaceuticals and all pharmaceuticals summed across three counties and three combinations of wastewater treatment scenario in 2050. *Source:* Welch et al. (2023) / OSF Preprints / CC BY 4.0.

a similar pattern in increases with the number of pharmaceuticals contributing to the indicator (not shown).

Future Outlooks

To the author's knowledge, the OOBN developed here is to date the most in-depth Bayesian model of pharmaceutical environmental risk, and the author hope its development, principally as a software demonstration, will serve as a foundation for the eventual development of useful tools for risk assessors and managers.

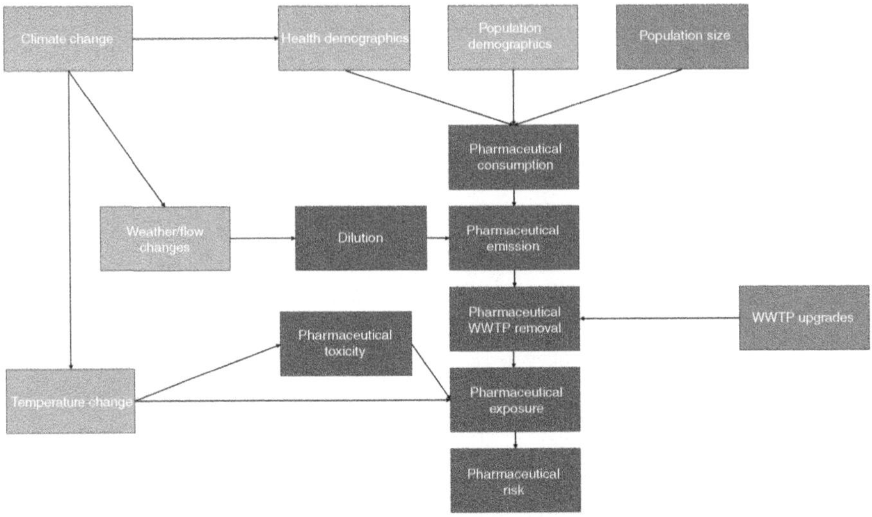

Figure 5.6 A possible future layout of an OOBN for predicting pharmaceutical risk in Norway.

The ability of this network to incorporate all relevant drivers of pharmaceutical risk is necessarily limited, due to the lack of information on quantitative (or even qualitative importance) of different drivers. Therefore, a key priority of future work is to conduct some sort of assessment of the most important drivers, at least in the Norwegian context, and restructure the network to allow for focus on these factors (Figure 5.6). Furthermore, if the costs of different mitigation and management options (e.g., WWTP upgrades) can be added as utility nodes (Barton et al. 2012), it would be possible to assess the costs and benefits of different approaches more holistically.

Given the criticism levied at RQs and their summation earlier in this chapter, it may have surprised the reader to see their use throughout this work as the primary measure of risk. In particular, the use of probability distributions of RQs is, on the face of it, an odd choice, as probability is being expressed twice – once, hidden, in the RQ, and a second time, transparently, as the probability distribution. This is clearly not ideal and was informed largely by a need to find a time-effective and comprehensible output of "risk." However, in the future the author would hope to use measures of environmental impact more anchored in real-world ecosystem protection and risk management goals. For instance, Landis et al. (2017) were able to relate various stressors to impact on various politically important end-points (human health and water use, population of key species), which has the added benefit of producing more measurable predictions than can be subsequently validated against real-world conditions.

Conclusions

In this work, we developed an OOBN to demonstrate how such approaches can be used to address the complex environmental challenge of pharmaceuticals in Norway. Although the initial network remains, by design, of limited scope and practical applicability, the approach yields a number of advantages for ERA (Moe et al. 2022), and it is hoped that with further work it can be shaped into a useful tool for the exploration, assessment, and management of environmental risks.

Acknowledgments

The author gratefully acknowledges Jannicke Moe, Merete Grung, and Sophie Mentzel of the Norwegian Institute for Water Research (NIVA) for their advice on this chapter, and Knut Erik Tollefsen of the NIVA Computational Toxicology Program for providing funding during its writing. This chapter is based on work that received funding from the EU Horizon 2020 program, under Marie Skłodowska-Curie grant agreement No 813124.

References

van der Aa, M. and Kommer, G. (2010). Forecast of pharmaceutical consumption in the Netherlands using demographic projections. In: *Green and Sustainable Pharmacy* (ed. K. Kümmerer and M. Hempel), 201–208. Berlin, Heidelberg: Springer. https://doi.org/10.1007/978-3-642-05199-9_13.

van der Aa, N.G.F.M., Kommer, G.J., van Montfoort, J.E., and Versteegh, J.F.M. (2011). Demographic projections of future pharmaceutical consumption in the Netherlands. *Water Sci. Technol. J. Int. Assoc. Water Pollut. Res.* 63 (4): 825–831. https://doi.org/10.2166/wst.2011.120.

Adams, K.J., Macleod, C.A.J., Metzger, M.J. et al. (2022). Developing a Bayesian network model for understanding river catchment resilience under future change scenarios. *EGUsphere* 1–35. https://doi.org/10.5194/egusphere-2022-617.

Allen, J.H. (2013). The wicked problem of chemicals policy: opportunities for innovation. *J. Environ. Stud. Sci.* 3 (2): 101–108. https://doi.org/10.1007/s13412-013-0117-0.

Backhaus, T. (2014). Medicines, shaken and stirred: a critical review on the ecotoxicology of pharmaceutical mixtures. *Philos. Trans. R. Soc. B, Biol. Sci.* 369 (1656): 20130585. https://doi.org/10.1098/rstb.2013.0585.

Bagath, M., Krishnan, G., Devaraj, C. et al. (2019). The impact of heat stress on the immune system in dairy cattle: a review. *Res. Vet. Sci.* 126: 94–102. https://doi.org/10.1016/j.rvsc.2019.08.011.

Barton, D.N., Kuikka, S., Varis, O. et al. (2012). Bayesian networks in environmental and resource management. *Integr. Environ. Assess. Manag.* 8 (3): 418–429. https://doi.org/10.1002/ieam.1327.

Bavumiragira, J.P., Ge, J., and Yi, H. (2022). Fate and transport of pharmaceuticals in water systems: a processes review. *Sci. Total Environ.* 823: 153635. https://doi.org/10.1016/j.scitotenv.2022.153635.

Bellard, C., Marino, C., and Courchamp, F. (2022). Ranking threats to biodiversity and why it doesn't matter. *Nat. Commun.* 13 (1): 2616. https://doi.org/10.1038/s41467-022-30339-y.

Bielen, A., Simatovic, A., Kosic-Vuksic, J. et al. (2017). Negative environmental impacts of antibiotic-contaminated effluents from pharmaceutical industries. *Water Res.* 126: 79–87. https://doi.org/10.1016/j.watres.2017.09.019.

Bolesta, W., Głodniok, M., and Styszko, K. (2022). From sewage sludge to the soil – transfer of pharmaceuticals: a review. *Int. J. Environ. Res. Public Health* 19 (16): 10246. https://doi.org/10.3390/ijerph191610246.

Bouzas-Monroy, A., Wilkinson, J.L., Melling, M., and Boxall, A.B.A. (2022). Assessment of the potential ecotoxicological effects of pharmaceuticals in the world's rivers. *Environ. Toxicol. Chem.* https://doi.org/10.1002/etc.5355.

Boxall, A.B.A., Rudd, M.A., Brooks, B.W. et al. (2012). Pharmaceuticals and personal care products in the environment: what are the big questions? *Environ. Health Perspect.* 120 (9): 1221–1229. https://doi.org/10.1289/ehp.1104477.

Bunke D, Moritz S, Brack W, Lopez-Herraez, Posthuma L, Nuss M. 2019. Developments in society and implications for emerging pollutants in the aquatic environment. *Environ. Sci. Eur.* 31, 32. https://doi.org/10.1186/s12302-019-0213-1.

Carmona, G., Varela-Ortega, C., and Bromley, J. (2011). The use of participatory object-oriented Bayesian networks and agro-economic models for groundwater management in Spain. *Water Resour. Manag.* 25 (5): 1509–1524. https://doi.org/10.1007/s11269-010-9757-y.

Chiabai, A., Quiroga, S., Martinez-Juarez, P. et al. (2018). The nexus between climate change, ecosystem services and human health: towards a conceptual framework. *Sci. Total Environ.* 635: 1191–1204. https://doi.org/10.1016/j.scitotenv.2018.03.323.

Cook, S.M., VanDuinen, B.J., Love, N.G., and Skerlos, S. (2012). Life cycle comparison of environmental emissions from three disposal options for unused pharmaceuticals. *Environ. Sci. Technol.* 46 (10): 5535–5541. https://doi.org/10.1021/es203987b.

Cook, B.I., Mankin, J.S., and Anchukaitis, K.J. (2018). Climate change and drought: from past to future. *Curr. Clim. Chang. Rep.* 4 (2): 164–179. https://doi.org/10.1007/s40641-018-0093-2.

Courtier, A., Cadiere, A., and Roig, B. (2019). Human pharmaceuticals: why and how to reduce their presence in the environment. *Curr. Opin. Green Sustain. Chem.* 15: 77–82. https://doi.org/10.1016/j.cogsc.2018.11.001.

Daughton, C.G. (2014). Eco-directed sustainable prescribing: feasibility for reducing water contamination by drugs. *Sci. Total Environ.* 493: 392–404. https://doi.org/10.1016/j.scitotenv.2014.06.013.

De Bruijn, J., Hansen, B.G., Johansson, S. et al. (2002). Technical guidance document on risk assessment. Parts 1–4. https://publications.jrc.ec.europa.eu/repository/handle/JRC23785 (accessed 19 September 2022).

Desrousseaux, A., Nagesh, P., Gajraj, R. et al. (2022). A shared socio-economic pathway based framework for characterising future emissions of chemicals to the natural environment. *Futures* 144: 103040. https://doi.org/10.1016/j.futures.2022.103040.

de' Donato, F. and Michelozzi, P. (2014). Climate change, extreme weather events and health effects. In: *The Mediterranean Sea: Its History and Present Challenges* (ed. S. Goffredo and Z. Dubinsky), 617–624. Dordrecht: Springer Netherlands. https://doi.org/10.1007/978-94-007-6704-1_38.

EurEau (2023). Position paper on the proposal for a directive concerning urban wastewater treatment (recast). https://www.eureau.org/resources/position-papers/7023-eureau-position-paper-on-the-proposal-for-a-directive-concerning-uwwtd/file (accessed 1 March 2023).

European Commission (2022). Proposal for a directive of the European parliament and of the council concerning urban wastewater treatment (recast). https://eur-lex.europa.eu/legal-content/EN/TXT/?uri=CELEX:52022PC0541 (accessed 14 February 2023).

Fairbrother, A., Hartl, B., Hope, B.K. et al. (2016). Risk management decisions for pesticides and threatened and endangered species: the role of uncertainty analysis. *Hum. Ecol. Risk Assess. Int. J.* 22 (2): 502–518. https://doi.org/10.1080/10807039.2015.1089400.

Fuller, R., Landrigan, P.J., Balakrishnan, K. et al. (2022). Pollution and health: a progress update. *Lancet Planet. Heath* 6 (6): e535–e547. https://doi.org/10.1016/S2542-5196(22)00090-0.

Gerbersdorf, S.U., Hollert, H., Brinkmann, M. et al. (2011). Anthropogenic pollutants affect ecosystem services of freshwater sediments: the need for a "triad plus x" approach. *J. Soils Sediments* 11 (6): 1099–1114. https://doi.org/10.1007/s11368-011-0373-0.

Godfrey, R. and Julien, M. (2005). Urbanisation and health. *Clin. Med.* 5 (2): 137–141. https://doi.org/10.7861/clinmedicine.5-2-137.

González Peña, O.I., López Zavala, M.Á., and Cabral Ruelas, H. (2021). Pharmaceuticals market, consumption trends and disease incidence are not driving the pharmaceutical research on water and wastewater. *Int. J. Environ. Res. Public Health* 18 (5): 2532. https://doi.org/10.3390/ijerph18052532.

Hart, B.T. and Pollino, C.A. (2008). Increased use of Bayesian network models will improve ecological risk assessments. *Hum. Ecol. Risk Assess. Int. J.* 14 (5): 851–853. https://doi.org/10.1080/10807030802235037.

Helwig, K., Niemi, L., Stenuick, J.-Y. et al. (2023). Broadening the perspective on reducing pharmaceutical residues in the environment. *Environ. Toxicol. Chem.* https://doi.org/10.1002/etc.5563.

Hope, B.K. (2009). Will there ever be a role for risk assessments? *Hum. Ecol. Risk Assess. Int. J.* 15 (1): 1–6. https://doi.org/10.1080/10807030802614819.

Ikehata, K. (2019). Recent research on ozonation by-products in water and wastewater treatment: formation, control, mitigation, and other relevant topics. In: *Water and Wastewater Treatment Technologies* (ed. X.-T. Bui, C. Chiemchaisri, T. Fujioka, and S. Varjani), 117–144. Singapore: Springer (Energy, Environment, and Sustainability). https://doi.org/10.1007/978-981-13-3259-3_7.

Inger, A. (2021). Speech prepared for delivery at public health from climate-related threats, on the margins of the 2021 united nations high-level political forum on sustainable development. 7 August. http://www.unep.org/news-and-stories/speech/prevention-versus-cure-climate-and-health-agendas (accessed 13 May 2023).

IPCC (2022). Summary for policymakers [H.-O. Pörtner, D.C. Roberts, E.S. Poloczanska, K. Mintenbeck, M. Tignor, A. Alegría, M. Craig, S. Langsdorf, S. Löschke, V. Möller, A. Okem (eds.)]. In: *Climate Change 2022: Impacts, Adaptation and Vulnerability. Contribution of Working Group II to the Sixth Assessment Report of the Intergovernmental Panel on Climate Change* (ed. H.-O. Pörtner, D.C. Roberts, M. Tignor, et al.), 3–33. Cambridge and New York: Cambridge University Press. https://doi.org/10.1017/9781009325844.001.

Jackson, M.C., Loewen, C.J.G., Vinebrooke, R.D., and Chimimba, C.T. (2016). Net effects of multiple stressors in freshwater ecosystems: a meta-analysis. *Glob. Chang. Biol.* 22 (1): 180–189. https://doi.org/10.1111/gcb.13028.

Jager, T., Vermeire, T.G., Rikken, M.G., and van der Poel, P. (2001). Opportunities for a probabilistic risk assessment of chemicals in the European Union. *Chemosphere* 43 (2): 257–264. https://doi.org/10.1016/S0045-6535(00)00087-4.

Jobling, S., Nola, M., Tyler, C.R. et al. (1998). Widespread sexual disruption in wild fish. *Environ. Sci. Technol.* 32 (17): 2498–2506. https://doi.org/10.1021/es9710870.

Koller, D. and Pfeffer, A. (1997). Object-oriented Bayesian networks. *Proceedings of the Thirteenth Annual Conference on Uncertainty in Artificial Intelligence*, Providence, Rhode Island, August 1-3, 1997.

Kong, D., MacLeod, M., Li, Z., and Cousins, I.T. (2013). Effects of input uncertainty and variability on the modelled environmental fate of organic pollutants under global climate change scenarios. *Chemosphere* 93 (9): 2086–2093. https://doi.org/10.1016/j.chemosphere.2013.07.049.

Landis, W.G. and Chapman, P.M. (2011). Well past time to stop using NOELs and LOELs. *Integr. Environ. Assess. Manag.* 7 (4): vi–viii. https://doi.org/10.1002/ieam.249.

Landis, W.G., Markiewicz, A., Ayre, K.K. et al. (2017). A general risk-based adaptive management scheme incorporating the Bayesian network relative risk model with the South River, Virginia, as case study. *Integr. Environ. Assess. Manag.* 13 (1): 115–126. https://doi.org/10.1002/ieam.1800.

Leeuwen, C.J.V. (2007). Introduction. In: *Risk Assessment of Chemicals: An Introduction* (ed. C.J. Leeuwen and Y.G. Vermeire), 1–36. Dordrecht: Springer Netherlands. https://doi.org/10.1007/978-1-4020-6102-8_1.

Lehikoinen, A. (2014). Bayesian network applications for environmental risk assessment. https://helda.helsinki.fi/items/ed120e8f-f682-4ef9-8709-92dd16c22055 (accessed 16 May 2023).

Lertxundi, U., Hernandez, R., Medrano, J., and Orive, G. (2020). Drug pollution and pharmacotherapy in psychiatry: a "platypus" in the room. *Eur. Psychiatry* 63 (1): e33. https://doi.org/10.1192/j.eurpsy.2020.32.

Li, Y., Zhu, G., Ng, W.J., and Tan, S.K. (2014). A review on removing pharmaceutical contaminants from wastewater by constructed wetlands: design, performance and mechanism. *Sci. Total Environ.* 468–469: 908–932. https://doi.org/10.1016/j.scitotenv.2013.09.018.

Li, Z., Sobek, A., and Radke, M. (2016). Fate of pharmaceuticals and their transformation products in four small European rivers receiving treated wastewater. *Environ. Sci. Technol.* 50 (11): 5614–5621. https://doi.org/10.1021/acs.est.5b06327.

Liang, W. and Yang, M. (2019). Urbanization, economic growth and environmental pollution: evidence from China. *Sustainable Comput. Inf. Syst.* 21: 1–9. https://doi.org/10.1016/j.suscom.2018.11.007.

Logar, I., Brouwer, R., Maurer, M., and Ort, C. (2014). Cost-benefit analysis of the Swiss national policy on reducing micropollutants in treated wastewater. *Environ. Sci. Technol.* 48 (21): 12500–12508. https://doi.org/10.1021/es502338j.

Longo, S., Mirko d' Antoni, B., Bongards, M. et al. (2016). Monitoring and diagnosis of energy consumption in wastewater treatment plants: a state of the art and proposals for improvement. *Appl. Energy* 179: 1251–1268. https://doi.org/10.1016/j.apenergy.2016.07.043.

Lundgren, K., Kuklane, K., Gao, C., and Holmer, I. (2013). Effects of heat stress on working populations when facing climate change. *Ind. Health* 51 (1): 3–15. https://doi.org/10.2486/indhealth.2012-0089.

Ma, J., Hung, H., and Macdonald, R.W. (2016). The influence of global climate change on the environmental fate of persistent organic pollutants: a review with emphasis on the Northern Hemisphere and the Arctic as a receptor. *Glob. Planet. Chang.* 146: 89–108. https://doi.org/10.1016/j.gloplacha.2016.09.011.

Malhi, Y., Franklin, J., Seddon, N. et al. (2020). Climate change and ecosystems: threats, opportunities and solutions. *Philos. Trans. R. Soc. B, Biol. Sci.* 375 (1794): 20190104. https://doi.org/10.1098/rstb.2019.0104.

Mangold-Döring, A., Huang, A., van Nes, E.H. et al. (2022). Explicit consideration of temperature improves predictions of toxicokinetic–toxicodynamic models for flupyradifurone and imidacloprid in *Gammarus pulex*. *Environ. Sci. Technol.* 56 (22): 15920–15929. https://doi.org/10.1021/acs.est.2c04085.

Massoud, M.A., Tarhini, A., and Nasr, J.A. (2009). Decentralized approaches to wastewater treatment and management: applicability in developing countries. *J. Environ. Manag.* 90 (1): 652–659. https://doi.org/10.1016/j.jenvman.2008.07.001.

McRae, D., Allman, M., and James, D. (2016). The redistribution of medicines: could it become a reality? *Int. J. Pharm. Pract.* 24 (6): 411–418. https://doi.org/10.1111/ijpp.12275.

Mitchell, C.J., Lawrence, E., Chu, V.R. et al. (2021). Integrating metapopulation dynamics into a Bayesian network relative risk model: assessing risk of pesticides to chinook salmon (*Oncorhynchus tshawytscha*) in an ecological context. *Integr. Environ. Assess. Manag.* 17 (1): 95–109. https://doi.org/10.1002/ieam.4357.

Mochon, S. (1993). When can you meaningfully add rates, ratios and fractions? *FLM* 13 (3): 16–21.

Moe, S.J., Benestad, R.E., and Landis, W.G. (2022). Robust risk assessments require probabilistic approaches. *Integr. Environ. Assess. Manag.* 18 (5): 1133–1134. https://doi.org/10.1002/ieam.4660.

Mora, C., McKenzie, T., Gaw, I.M. et al. (2022). Over half of known human pathogenic diseases can be aggravated by climate change. *Nat. Clim. Chang.* 12 (9): 869–875. https://doi.org/10.1038/s41558-022-01426-1.

Muñoz, I., Lopez-Doval, J.C., Ricart, M. et al. (2009). Bridging levels of pharmaceuticals in river water with biological community structure in the Llobregat river basin (northeast Spain). *Environ. Toxicol. Chem.* 28 (12): 2706–2714. https://doi.org/10.1897/08-486.1.

Myers, S.S. and Patz, J.A. (2009). Emerging threats to human health from global environmental change. *Annu. Rev. Environ. Resour.* 34 (1): 223–252. https://doi.org/10.1146/annurev.environ.033108.102650.

NICE Medicines and Prescribing Centre (UK) (2015). *Medicines Optimisation: The Safe and Effective Use of Medicines to Enable the Best Possible Outcomes*. Manchester: National Institute for Health and Care Excellence (NICE). http://www.ncbi.nlm.nih.gov/books/NBK305021/ (accessed 16 May 2023).

Norsk Vann (2023). Forslag til revidert avløpsdirektiv – Norsk Vanns posisjonsdokument. https://norskvann.no/wp-content/uploads/Norsk-Vanns-posisjonsdokument-kortversjon-20.2.23-2.pdf (accessed 1 March 2023).

O'Neill, B.C., Kriegler, E., Riahi, K. et al. (2014). A new scenario framework for climate change research: the concept of shared socioeconomic pathways. *Clim. Chang.* 122 (3): 387–400. https://doi.org/10.1007/s10584-013-0905-2.

Ormerod, S.J., Dobson, M., Hildrew, A.G., and Townsend, C.R. (2010). Multiple stressors in freshwater ecosystems. *Freshw. Biol.* 55 (s1): 1–4. https://doi.org/10.1111/j.1365-2427.2009.02395.x.

Osorio, V., Proia, L., Ricart, M. et al. (2014). Hydrological variation modulates pharmaceutical levels and biofilm responses in a Mediterranean river. *Sci. Total Environ.* 472: 1052–1061. https://doi.org/10.1016/j.scitotenv.2013.11.069.

Osuhor, P.C. and Essien, E.S. (1978). Urbanisation and its effects on environmental sanitation. *Public Health* 92 (2): 69–75. https://doi.org/10.1016/S0033-3506(78)80031-6.

Persson, L., Almroth, B.M.C., Collins, C.D. et al. (2022). Outside the safe operating space of the planetary boundary for novel entities. *Environ. Sci. Technol.* 56 (3): 1510–1521. https://doi.org/10.1021/acs.est.1c04158.

Pirotta, E., Thomas, L., Costa, D.P. et al. (2022). Understanding the combined effects of multiple stressors: a new perspective on a longstanding challenge. *Sci. Total Environ.* 821: 153322. https://doi.org/10.1016/j.scitotenv.2022.153322.

Raimondo, S. and Forbes, V.E. (2022). Moving beyond risk quotients: advancing ecological risk assessment to reflect better, more robust and relevant methods'. *Ecologies* 3 (2): 145–160. https://doi.org/10.3390/ecologies3020012.

Rockström, J., Stefen, W., Noone, K. et al. (2009). Planetary boundaries: exploring the safe operating space for humanity. *Ecol. Soc.* 14 (2): art32. https://doi.org/10.5751/ES-03180-140232.

Shaban, A., Kourtit, K., and Nijkamp, P. (2022). Causality between urbanization and economic growth: evidence from the Indian states. *Front. Sustainable Cities* 4. https://www.frontiersin.org/articles/10.3389/frsc.2022.901346 (accessed 13 May 2023).

Sommerschild, H.T., Berg, C.L., Jonasson, C. et al. (2021). Data resource profile: Norwegian databases for drug utilization and pharmacoepidemiology. *Nor. Epidemiol.* 29 (1–2): https://doi.org/10.5324/nje.v29i1-2.4040.

Steffen, W., Richardson, K., Rockstrom, J. et al. (2015). Planetary boundaries: guiding human development on a changing planet. *Science* 347 (6223): 1259855. https://doi.org/10.1126/science.1259855.

Sumpter, J.P., Johnson, A.C., and Runnalls, T.J. (2022). Pharmaceuticals in the aquatic environment: no answers yet to the major questions. *Environ. Toxicol. Chem.* https://doi.org/10.1002/etc.5421.

Swan, G.E., Cuthbert, R., Quevedo, M. et al. (2006). Toxicity of diclofenac to Gyps vultures. *Biol. Lett.* 2 (2): 279–282. https://doi.org/10.1098/rsbl.2005.0425.

Swedish Chemicals Agency (2021). Improving the regulatory assessment of combination effects: steps towards implementing the mixture assessment factor (MAF) in chemical regulation. *Report PM* 8 (21): 61.

Thomas, M.J. and Tømmerås, A.M. (2022). Norway's 2022 national population projections, p. 158.

Topaz, T., Boxall, A., Suari, Y. et al. (2020). Ecological risk dynamics of pharmaceuticals in micro-estuary environments'. *Environ. Sci. Technol.* 54 (18): 11182–11190. https://doi.org/10.1021/acs.est.0c02434.

UN Department of Economic and Social Affairs, Population Division (2022). World population prospects 2022: summary of results. UN DESA/POP/2022/TR/NO. 3. https://www.un.org/development/desa/pd/sites/www.un.org.development.desa.pd/files/wpp2022_summary_of_results.pdf (accessed 31 March 2023).

Van Boeckel, T.P., Brower, C., Gilbert, M. et al. (2015). Global trends in antimicrobial use in food animals. *Proc. Natl. Acad. Sci.* 112 (18): 5649–5654. https://doi.org/10.1073/pnas.1503141112.

Van Boeckel, T.P., Pires, J., Silvester, R. et al. (2019). Global trends in antimicrobial resistance in animals in low- and middle-income countries. *Science* 365 (6459): eaaw1944. https://doi.org/10.1126/science.aaw1944.

Vatovec, C., Van Wagoner, E., and Evans, C. (2017). Investigating sources of pharmaceutical pollution: survey of over-the-counter and prescription medication purchasing, use, and disposal practices among university students. *J. Environ. Manag.* 198: 348–352. https://doi.org/10.1016/j.jenvman.2017.04.101.

Verlicchi, P. and Zambello, E. (2014). How efficient are constructed wetlands in removing pharmaceuticals from untreated and treated urban wastewaters? A review. *Sci. Total Environ.* 470–471: 1281–1306. https://doi.org/10.1016/j.scitotenv.2013.10.085.

Vos, T., Lim, S.S., Abbafati, C. et al. (2020). Global burden of 369 diseases and injuries in 204 countries and territories, 1990–2019: a systematic analysis for the global burden of disease study 2019'. *Lancet* 396 (10258): 1204–1222. https://doi.org/10.1016/S0140-6736(20)30925-9.

van Vuuren, D.P., Riahi, K., Moss, R. et al. (2012). A proposal for a new scenario framework to support research and assessment in different climate research communities. *Glob. Environ. Chang.* 22 (1): 21–35. https://doi.org/10.1016/j.gloenvcha.2011.08.002.

Wang, H., Yang, Y., Keller, A.A. et al. (2016). Comparative analysis of energy intensity and carbon emissions in wastewater treatment in USA, Germany, China and South Africa. *Appl. Energy* 184: 873–881. https://doi.org/10.1016/j.apenergy.2016.07.061.

Wang-Erlandsson, L., Tobian, A., van der Ent, R. et al. (2022). A planetary boundary for green water. *Nat. Rev. Earth Environ.* 3 (6): 380–392. https://doi.org/10.1038/s43017-022-00287-8.

Welch, S.A., Grung, M., Madsen, A.L., and Moe, J. (2023). Probabilistic risk calculation for chemical mixtures: environmental risk of pharmaceuticals under future scenarios. OSF Preprints. https://doi.org/10.31219/osf.io/zbgp7

WHO (2021). Climate change and health. https://www.who.int/news-room/fact-sheets/detail/climate-change-and-health (accessed 27 February 2023).

WHO (2022). A toolkit on how to implement social prescribing. https://www.who.int/publications-detail-redirect/9789290619765 (accessed 16 May 2023).

WHO and CBD (2015). *Connecting Global Priorities: Biodiversity and Human Health: A State of Knowledge Review*. Geneva: World Health Organization. https://apps.who.int/iris/handle/10665/174012 (accessed 21 March 2023).

Wilkinson, J.L., Boxall, A.B.A., Kolpin, D. et al. (2022). Pharmaceutical pollution of the world's rivers. *Proc. Natl. Acad. Sci.* 119 (8): e2113947119. https://doi.org/10.1073/pnas.2113947119.

World Economic Forum (2023). Global risks report 2023, world economic forum. https://www.weforum.org/reports/global-risks-report-2023/in-full/1-global-risks-2023-today-s-crisis/ (accessed 27 February 2023).

Yu, W.G.L., Sofield, R.M., and Ho, M. (2017). Ecological risk assessment. In: *Introduction to Environmental Toxicology*, 5e. CRC Press.

6

The Role of Contaminants, in the Context of Multiple Stressors, in the Collapse of a Striped Bass Population

David J. Ostrach[1] and Cameron Irvine[2]

[1]Chief Scientist at Ostrach Consulting, Woodland, CA, USA
[2]Associate Environmental Scientist at Robertson-Bryan, Inc., Environment Department, Rancho Cordova, CA, USA

Introduction

Abundances of pelagic fish species in the Sacramento-San Joaquin River Delta and San Francisco Bay Estuary (Bay-Delta) have declined since the 1970s. Although these declines have been, in part, consistent with natural long-term decadal-scale oscillations, certain features suggest additional anthropogenic-induced mechanisms. These unusual features include a possible steplike decline near the turn of the century and a failure of abundance measures to favorably respond to recent years of high precipitation and Delta inflows. Possible anthropogenic-induced mechanisms, as outlined in the Pelagic Organism Decline (POD) conceptual model, include (1) food web interactions, (2) toxicity, and (3) water project operations (Baxter et al. 2008; Sommer et al. 2011). A summary of research in the Bay-Delta associated with striped bass that was among several studies conducted to investigate the multiple-stressors contributing to the POD is provided in the forthcoming section. This chapter summarizes a technical report prepared by Dr. David Ostrach for the Department of Water Resources (Ostrach 2009). The original technical report presents detailed data and methodological descriptions (https://www.waterboards.ca.gov/waterrights/water_issues/programs/bay_delta/docs/cmnt091412/sldmwa/ostrach_et_al_2009.pdf).

Approach

The factors and associated interactions that contributed and continue to contribute to the decline of pelagic fish in the Bay-Delta are admittedly complicated and multifaceted. Understanding issues of these types generally requires a breadth of analytical techniques and a highly interdisciplinary team of investigators. This study of striped bass in the Bay-Delta assembled scientists with broad scientific backgrounds that included expertise in histopathology, analytical chemistry, geochemistry, toxicology, protein profiling, ecological modeling, spatial mapping, microbiology, and parasitology.

An underlying hypothesis was that interactions of multiple stressors in the Bay-Delta region substantially contributed to the decline of the striped bass population, specifically that (1) sublethal exposures of early life stages of striped bass to contaminants induces pathological and physiological changes that adversely affect subsequent survival and ultimately population numbers (Okihiro et al. 1990; Bailey et al. 1991; Bailey et al. 1993; Bennett et al. 1995; Ostrach 2006; Ostrach et al. 2008); (2) regional contaminant interactions over a broad area and diverse history of uses is best accomplished by addressing contaminants as a unit, rather than a list, to uncover indirect as well as direct toxic effects such as depletion of effective immune response mechanisms and reproductive efficacy while addressing the interactions of multiple stressors (Springman et al. 2005); and (3) health effects associated with the interaction of multiple stressors (i.e., sublethal exposure to contaminants, food limitation, and disease) in the Bay-Delta substantially contributed to the decline of the striped bass population (Ostrach 2009). The significance of contaminants on the decline of the striped bass population was one of several studies investigating factors in the POD conceptual model (Baxter et al. 2008). The overall objectives in this study was to assess the health status of larval, juvenile, and adult female striped bass collected from selected locations in the Bay-Delta using morphometric, histopathological, otolith (aging, growth, and microgeochemical analyses) and biochemical metrics.

Contaminant Exposures

Contaminant data is abundant in the Bay-Delta, and concentrations of organic compounds and mercury can exceed water quality benchmarks or screening values (Davis et al. 2000; de Parsia et al. 2018). Mercury concentrations in striped bass from the Bay-Delta frequently exceed the screening value of 0.24 mg/kg wet weight in fillets of trophic level 4 fish (i.e., fish that eat other fish) established by the Central Valley Regional Water Quality Control Board (2010). Likewise, polychlorinated biphenyls (PCBs) and dichlorodiphenyltrichloroethane (DDT) in largemouth bass and white catfish from the Bay-Delta are elevated in local hot spots (de Parsia et al. 2018). Other chemicals of concern include dieldrin, toxaphene, arsenic, polycyclic aromatic hydrocarbons (PAHs), dioxins, and pesticides.

Tissue burdens of these chemicals can improve our understanding of the role habitat residence plays in the exposure, bioaccumulation, and maternal transfer of xenobiotics in Bay-Delta fishes. Polybrominated diphenyl ether (PBDE), PCB, DDT, DDT degradation compounds, and various other legacy contaminant concentrations in striped bass eggs collected in 1990, 1992, 1999, 2001, and 2006 were sufficiently high to adversely affect larval striped bass (Ostrach et al. 2008; Ostrach et al. 2009). Concentrations were also associated with habitat use and sediment concentrations. A weak-positive correlation was found between Delta residence and PBDE levels but not between PBDE accumulation and residence time of striped bass between Carquinez Strait and the Pacific Ocean. PBDE bioaccumulation was also positively correlated with fork length in female striped bass. This finding suggested that female striped bass do not maternally transfer the entire PBDE load to their eggs but retain and continue to bioaccumulate higher levels as they age and increase in size. Thus, the metabolism of PBDE in striped bass and maternal transfer to the eggs is different in this class of compounds than in other lipophilic contaminants such as PCBs. There were no correlations between PCB concentrations and Bay-Delta residence, or with PCBs and fish age or fork length. Furthermore, a slightly positive correlation was found between PCB bioaccumulation and residence time of striped bass between Carquinez Strait and the Pacific Ocean. These results are consistent with sediment sampling where PCB concentrations were higher in San Francisco Bay than in the Delta, and PBDE levels in the Delta were higher than in San Francisco Bay. Such data help resource managers consider the importance of striped bass migration, habitat use, and striped bass size to understand how migration patterns subject these fish to the greatest risk of exposure to xenobiotics and when determining safe consumption limits for these fish.

Biochemical Activation

Biochemical and molecular methods have been developed for monitoring environmental exposures to environmental toxicants in striped bass. These biomarkers allow for tissue-specific sampling that can focus investigations on the site of toxic action – where chemicals exert their effect(s). While activation of some biomarkers signal a response to exposure to specific chemicals, others are linked with apical (i.e., whole-organism) effects that have been associated with survival, growth, reproduction, and even population-level effects. Applicable endpoints include changes in liver and other tissue enzymes involved in metabolizing environmental toxicants, as well as levels of proteins with reproductive and protective functions induced by exposure to environmental toxicants (Heppell et al. 1995; Hodson et al. 1996; George et al. 2004; Sarkar et al. 2006). Contaminant-based mechanisms involved in the abnormal growth and development observed in wild-caught striped bass larvae were assessed in the Sacramento-San Joaquin Delta using EROD and AChE, vitellogenin, and metallothionein biochemical assays (Ostrach et al. 2009; Durieux et al. 2011; Spearow et al. 2011).

Interactions among physiological systems when organisms are exposed to multiple types of contaminant exposure is complicated and can lead to an underestimation of exposure due to antagonism (i.e., when two chemical exposures cause the opposite biochemical response). Therefore, findings of positive single and multiple biomarkers are a conservative estimation, if not an underestimation, of the percentage of the juvenile striped bass population that were under sublethal contaminant exposure in the San Francisco Estuary and likely causing physiological stress and immunosuppression.

EROD

Changes in microsomal 7-ethoxyresorufin-O-deethylase (EROD) liver protein is a quantitative measure of contaminant exposure. Samples from juvenile striped bass in the Bay-Delta indicated that the majority of fish (114 of 178; ~65%) collected in August 2007 through January 2008 were under EROD induction. This was the third consecutive year in which EROD results indicated that the vast majority of juvenile striped bass were under significant sublethal contaminant exposures throughout their entire range in the first six months of life (Ostrach et al. 2009).

EROD activity is a commonly used biomarker for fish exposure to a broad array of toxicants, including polychlorinated dibenzo-p-dioxins (PCDDs), PCBs, polychlorinated dibenzofurans (PCDFs), PAHs, and certain insecticides that induce the CYP1A enzyme (Krüner and Westernhagen 1999; Whyte et al. 2000; Miller et al. 2004). The EROD assay measures the induction of cytochrome P4501A (CYP1A) and other cytochrome P450 enzymes, which are critical for metabolizing PAHs (Wassenberg et al. 2005; Billiard et al. 2006). However, induction of CYP1A/EROD activity can also be inhibited by environmentally relevant compounds, modulated by estrogenic agents, glucocorticoids, toxic metals, and seasonality (Krüner and Westernhagen 1999). CYP1A also activates these toxicants to their ultimate mutagenic/carcinogenic forms, and increases levels of reactive oxygen species (ROS), oxidative stress, apoptosis, and early developmental toxicity (Arzuaga et al. 2006). Thus, developing fish embryos and early life forms are very sensitive to CYP1A-associated oxidative stress and toxicity (Guiney et al. 1997; Wassenberg and Di Giulio 2004; Billiard et al. 2006; Voelker et al. 2007).

Exposure to CYP1A antagonists in the environment and in body tissues may inhibit the induction or activity of CYP1A in fish and their first-generation offspring (Meyer et al. 2002). The induction of EROD by TCDD was potently inhibited by several PBDE congeners (Kuiper et al. 2004). EROD activity can also be inhibited by exposure of fish to mercury and cadmium (Bozcaarmutlu and Arinc 2004) and to ketoconazole antifungals (Hegelund et al. 2004). Even common environmental and laboratory surfactants, including TritonX-100 and Tween 80, and organic solvent, methanol, markedly inhibits EROD activity (Krüner and Westernhagen 1999; da Silva and Meirelles 2004). These and other observations suggest that environmentally relevant body burdens may inhibit the induction of CYP1A by PAHs, thus potentiating PAH toxicity. In contrast, the induction of CYP1A, while adaptive to PAH exposures, can also enhance toxicity following exposure to dioxins and PCBs (Incardona et al. 2005; Wassenberg et al. 2005; Billiard et al. 2006).

Immunointegrity

Immunohistochemical (IHC) analysis for cytochrome P450 expression using the CYP1-A antibody was performed on subsets of fish from juvenile striped bass in the Bay-Delta (Ostrach et al. 2009). Analyses were conducted on multiple tissues (i.e., multifocal staining of the hepatopancreatic vessels, heart, and branchial lamellar vessels or the lamellar pillar cells). These IHC analyses indicated that the vast majority of juvenile striped bass throughout their entire range and first six months of life were under sublethal contaminant exposure as measured by cytochrome P450-1A1 expression for the third year in a row. This type of sublethal contaminant exposure causes physiological stress and likely immunosuppression, which may explain the abnormal findings of parasitism and disease in these juvenile striped bass.

Acetylcholinesterase Inhibition

Another toxicant-specific biomarker is the measurement of acetylcholinesterase (AChE) activity in response to exposure to organophosphate (OP) and carbamate insecticides. AChE is one of the two broad classes of choline esterases and is responsible for the removal of the neurotransmitter acetylcholine (ACh) from the synaptic cleft by hydrolyzing ACh into choline and acetic acid (Fulton and Key 2001). The irreversible binding of OPs and reversible binding of carbamates to the esteratic site of AChE potentiates the cholinergic effect, disrupting neurotransmitter processes. For vertebrates, these enzymes are vital for the somatic, parasympathetic, sympathetic, and central nervous systems (Fulton and Key 2001).

Ellman et al. (1961) developed a relatively simple method for quantifying AChE activity in vertebrate tissues, and numerous studies have shown that exposure to OPs or carbamates results in a concentration-dependent inhibition of AChE activity in various tissues (Mayer et al. 1992). Thus, AChE inhibition is linked directly with the mechanism of toxic action. AChE activity is one of the most common biomarkers of neurotoxicity used in aquatic organisms. One of the advantages of this method is that measuring significant AChE inhibition provides evidence that a sufficient dose of compound has reached the target site to produce a physiological effect. In addition, enzyme inhibition often persists for an extended period of time, while OP insecticide detection in the environment may prove unsuccessful due to rapid degradation (Fulton and Key 2001). AChE activity has been measured in various tissues of several fish species, and a number of studies have examined the relationship between specific levels of OP-induced AChE inhibition and lethality. Relatively high OP concentrations, such as those resulting from an accidental spills, cause hyperactivity, muscle twitching, loss of equilibrium, and ultimately death (Zinkl et al. 1991). However, environmental exposures encountered by fish are typically much lower, and studies have investigated the relationship between OP exposure, AChE inhibition, and sublethal effects. Sandahl et al. (2005) for example, specifically related observed changes in fish behavior to AChE inhibition in brain and muscle, thus linking the degree of enzyme inhibition to behavioral impairments such as compromised predator avoidance, territory defense, and reduced schooling and feeding success, which may, in turn, reduce survival or reproductive success.

AChE is also a means of investigating biological effects of complex mixtures of many neurotoxic contaminants on aquatic environments (Lionetto et al. 2005). The maternal transfer of contaminants and effects on embryonic and larval striped bass indicated that chemicals that could affect nervous system development and AChE levels are being transferred in biologically relevant levels from wild female striped bass (Ostrach 2006; Ostrach et al. 2008). Changes in AChE activity during early larval development of hatchery (control) versus river females' progeny was assessed in 2007 to test for potential effects caused by maternal transfer of contaminants. These developmental studies of larval striped bass AChE activity displayed a strong and significant increase throughout the developmental period. Wild-spawned larvae exhibited a threefold increase in AChE activity between day 1 and day 5 post-hatching. This increase correlated with the ongoing development of striped bass larvae's nervous and organ systems. Between day 1 and day 3 post-hatch, organogenesis is ongoing and includes development of the eyes, mouth, gills, axial musculature, fins, and the nervous system. The nervous system is also growing rapidly during this period to innervate musculature and sensory functions in the larvae. Findings of the threefold increase of activity in early life-stage striped bass demonstrated the importance of monitoring AChE activity in developing larvae. AChE activity and striped bass is strongly positively correlated with water temperature (Ostrach 2009). This environmental variable, along with electrical conductivity and for size, should be considering when investigating AChE as a neurotoxicity biomarker.

Wild-caught juvenile striped bass did not have a significant inhibition of AChE activity. This was not an unexpected finding as AChE inhibition occurs when fish are being exposed to significant levels of pesticides that did not coincide with sample collection locations. Thus, unlike adults, these juvenile striped bass data indicated that OP and carbamate pesticides were not being bioaccumulated through the water or food. However, adults use food sources and habitat that differ from juvenile fish, and the data are not inconsistent. Adult striped bass bioaccumulate significant chemical body burdens of legacy and current-use pesticides, and maternal transfer studies found that larval fish receive significant exposures to these chemicals in the egg (Ostrach 2006; Ostrach et al. 2008).

Metallothionein

Metallothionein (MT) is induced by metals including zinc, cadmium, mercury, and other chemical factors (Schlenk et al. 1995; Lacorn et al. 2001; Chung et al. 2005), and by severe oxidative stress (Viarengo et al. 1999; Chung et al. 2005). It is a heat-stable, evolutionary conserved, low-molecular-weight protein with a high proportion of cysteine residues. Activation of MT has been implicated in diverse biological functions including metal ion homeostasis, binding and regulating the normal physiological levels of zinc and copper, while also protecting against toxic metals including cadmium and mercury (Cai et al. 2000; Lacorn et al. 2001). The sequestering of toxic metals by MT markedly decreases their acute toxicity. More recent studies have shown that MT also protects against oxidative stress (Viarengo et al. 1999; Chung et al. 2005). Exposure to ROS and reactive nitrogen species (RNS) results in DNA strand breaks and increased apoptosis. As an antioxidant, MT functions to protect against oxidative damage of DNA, protein and lipids by RNS, and ROS including hydrogen peroxide, superoxide anion, and hydroxyl radical (Cai et al. 2000).

Exposure to toxic metals is more closely related to MT levels in the liver than in gill or kidney (Van Campenhout et al. 2003). Thus, we examined liver MT levels as a biomarker of the exposure of striped bass to toxic metal and oxidative stress-inducting environmental toxicants. In this study, expression of MT was at levels reported to indicate severe physiological stress, regardless of whether due to metals or oxidative stress alone (Ostrach 2009).

Vitellogenin

Studies around the world have found that reproductive development of fish is being disrupted by environmental exposures to endocrine-disrupting chemicals

(EDC) and cellular expression of genes or the protein vitellogenin (VTG) indicates exposure to EDCs. These chemicals enter surface waters from municipal and industrial effluents, municipal stormwater, and runoff from agriculture where pesticides have been applied. Exposure to estrogenic compounds in surface waters is, therefore, widespread and from many different compounds including 17-alpha ethinylestradiol, steroids, alkylphenols, phthalates, pyrethroids, and the antimicrobial chemical triclosan. EDC effects have also been widely documented for many reproductive endpoints, such as feminization in males, inducing intersexes/ovotestes, decreased sperm counts, decreased fertility, reduced breeding success (egg viability, hatchability, and swim-up success), reduced growth, as well as altered development and behavior (Harris et al. 2001; Thorpe et al. 2001; Ackermann et al. 2002; Jobling et al. 2002; Kang et al. 2003; Arsenault et al. 2004; Ferreira et al. 2004; Ishibashi et al. 2004; Fenske et al. 2005; Ishibashi et al. 2005; McCormick et al. 2005; Cionna et al. 2006; Mortensen et al. 2006; Sakamoto and McCormick 2006). These EDC effects have even resulted in population declines for several piscine species (van Aerle et al. 2002; Hill and Janz 2003; Jobling et al. 2006; Kidd et al. 2007).

To further complicate matters, several studies have shown that estrogenic and EROD-inducing chemicals can interact. Estrogenic agents including EE2 and 4-NP inhibit the induction of EROD activity by CYP1A agonists in European flounder and grey mullet (Vaccaro et al. 2005; Cionna et al. 2006; Kirby et al. 2007). The inhibition CYP1A induction occurs at much lower concentrations of estrogenic agents than required to induce VTG in European flounder and grey mullet (Cionna et al. 2006; Kirby et al. 2007). Furthermore, induction of EROD by PAHs has also been shown to inhibit the induction of VTG by E2 in rainbow trout hepatocytes (Navas and Segner 2000). Thus, the true levels of estrogens in PAH contaminated sites may be underestimated due to high levels of CYP1A/EROD activity inhibiting estrogenic responses. While estrogenic compounds can act directly through estrogen receptors to induce VTG levels, other chemicals can alter VTG and other estrogenic responses by affecting estrogen receptor levels, aromatase levels, or metabolism of estrogens. For example, tributyltin decreases VTG apparently by decreasing E2 receptor alpha (Mortensen and Arukwe 2007).

Hydroxylated PCBs can also be estrogenic, at least in part by inhibiting estrogen sulfotransferase and therefore the metabolism of estrogens (Kester et al. 2002). Thus, hydroxylated PCBs are likely to induce both EROD and VTG. These observations show significant cross-talk between the induction of EROD by the aryl hydrocarbon receptor pathway and the induction of VTG and reproduction by the estrogen receptor pathway. Thus, monitoring EROD activity alone may underestimate effects of environmental agents on the induction of CYP1A and EROD activity in watersheds containing estrogenic contaminants. Genetic differences in sensitivity to estrogenic agents are found in diverse species ranging from fish to mammals (Spearow et al. 1999). For example, sunshine bass were much more sensitive than mummichogs to induction of VTG by New York sewage effluent (McArdle et al. 2000). Thus, species and populations differences in sensitivity to EDCs need to be considered when investigating these effects. In the Bay-Delta, expression of VTG expression was elevated in a significant number of juvenile striped bass (Ostrach et al. 2009).

Biochemical Effects Synthesis

Metallothionein and VTG assays for juvenile striped bass indicated that a substantial number of fish were expressing MT (52 of 162 fish; 34%) and VTG (39 of 174 fish; 22%) in significantly elevated levels, and expression of these biomarkers was both spatially and temporally correlated. These findings, coupled with the IHC and EROD results indicated that juvenile striped bass in the Bay-Delta were under several types of sublethal contaminant exposure (Ostrach et al. 2009). In fish evaluated for multiple biomarkers of contaminant exposure, numerous (i.e., 18) fish were positive for both EROD and metallothionein, some (n = 17) were positive for EROD and VTG, others (n = 6) for VTG and MT, or positive for EROD, VTG, and MT (n = 13). These results indicated that a significant proportion of the juvenile striped bass population (i.e., 54 of 162 samples or 33%) were experiencing multiple sublethal contaminant exposures (Ostrach et al. 2009).

Interactions between the physiological systems involved with multiple contaminant exposures is complex and can lead to an underestimation of exposures. Therefore, these findings of positive single and multiple biomarkers were a conservative estimate, if not underestimate, of the percentage of the juvenile striped bass population that are under sublethal contaminant exposure in the Bay-Delta causing physiological stress and likely immunosuppression.

Egg-Ovaries

Adult striped bass from the Sacramento River, Mokelumne River, and San Joaquin River appeared healthy and had few observed gross lesions or internal/external parasites at the time of capture during surveys 1999–2001 and 2006–2007. However, concentrations of PCBs, PBDEs, and pesticides in eggs from wild fish in 1999–2001 and 2006 were elevated. Hatchery-reared eggs contained far fewer contaminants and at lower levels than those in the wild. DDT concentrations decreased from 1999–2001 to 2006. These detected concentrations in eggs demonstrate the significant maternal transfer of xenobiotics from female striped bass into their eggs (Ostrach et al. 2008; Ostrach 2009).

A few egg samples from females in the Mokelumne River had contaminant concentrations about half of those seen in fish from the Sacramento River. Habitat-use patterns of these fish, evaluated through analysis of otolith geochemistry, determined that the fish with very low levels of contaminants were residents that lived their entire life in the Mokelumne River; whereas striped bass with elevated concentrations of xenobiotics were nonresident fish using Delta habitats in a manner similar to the Sacramento River–collected females. This difference between the chemical concentrations of eggs from resident and nonresident fish may be due to differing diets of the individual fish as well as spatial variability in chemical exposures.

Analysis of the 2006 egg samples, coupled with habitat-use profiles of the females, indicated that fecundity estimates of striped bass should be reconsidered. All of the female striped bass from the Sacramento River, San Joaquin River, and some Mokelumne River had high levels of contaminants in their eggs that would affect larval development and survival. Therefore, traditional measures of fecundity (eggs/female = egg/larval production) do not correctly estimate larval striped bass production. Larvae from resident Mokelumne River fish could have a higher probability of developing normally and higher survival than larvae produced from females' exploiting other Delta habitats with higher chemical exposures and could be contributing relatively higher recruitment to the striped bass population. Nevertheless, the Mokelumne River is not a preferred spawning habitat for striped bass.

Developmental and Histopathology Effects

Larvae from eggs collected in the wild grew more slowly than hatchery-reared fish that were fed the same diets. Utilization of contaminated yolk in wild larvae caused abnormal organ development and growth. Livers of wild larvae grew rapidly and then shrunk, exhibiting lesions of contaminant exposure before first feeding (Ostrach et al. 2008). Growth of wild larvae ceased after three to five days, and these larvae had significant lesions not present in hatchery larvae. The overwhelming majority of these wild larvae died before first feeding. Developmental abnormalities, gross and histological lesions, disease, parasite and viral load, and biochemical markers were also observed in the progeny of striped bass larvae from river-collected females in 2006 and 2007 (Ostrach 2009) and were consistent with effects and associated with the maternal transfer of xenobiotics in 1999 and 2001 (Ostrach 2006; Ostrach et al. 2008). Gross lesions were found in 95% of the larvae spawned from wild-collected eggs while no lesions occurred in progeny from hatchery female striped bass. Histopathological analysis confirmed these gross lesion observations with abdominal edema, brain, and finfold edema; necrosis of epithelium; and necrotic lesions in the midbrain undifferentiated neuroblasts and glial cells. Lesions in the brain of developing larvae from river-collected females was consistent with numerous studies associated with xenobiotic exposures (Okihiro et al. 1990; Bailey et al. 1991; Bailey et al. 1994; Ostrach 2006; Ostrach et al. 2008; Ostrach 2009). These growth and developmental effects indicated serious problems with wild striped bass progeny in 2006. A combination of poor growth, abnormal development, and lesions suggested that very few wild larvae would survive to the juvenile stage, thus significantly reducing recruitment and adult population levels.

Histopathological effects in juvenile striped bass sampled in 2005–2007 were similar. The most common findings included external protozoan parasitic infections that primarily involved the branchial tissues (also the integument, oropharynx, and opercular cavities) and included *Trichodina* sp. coelomic granulomatous foci that were presumed to be due to this trematode infection and sessile ciliated protozoans. The most significant finding was the branchial protozoan infection in 65% of juvenile striped bass examined that was not only consistent with a trichodinid infection but also included other ciliated protozoan such as sessile ciliates. The level and severity of the infections were abnormally high for striped bass and for juvenile fish in general. Gill parasitism has been shown to adversely affect fish growth and interfere with respiration, especially during the first year or two of life (Sadzikowski and Wallace 1974). Thus, the level and severity of inflammation in the fish from 2005 with trematodiasis and the external protozoan infections in fish from 2006 and 2007 were a significant impact on fish health that could have compromised fish health, especially that of juvenile fish, and may have even led to mortalities. Viral isolation performed on selected fish in 2007 did not identify any viral agents or cytopathological effects that would have been caused by viruses.

It is important to note that antemortem severity of external parasitic infections cannot be definitively determined by histological examination since external parasitic agents will generally leave the host following death of the host or will be removed from the tissue following fixation of the tissues. As such, this finding is an underestimate of the actual level of infection in sampled wild fish. In this context, a more definitive determination of the severity of infection can only be determined by the cytological examination of branchial preparations using branchial tissue obtained from live fish or fish immediately following euthanasia. In addition, a definitive identification of the protozoan parasites cannot be performed on histological sections, and it also requires cytological preparations. However, *Trichodiniasis* can result in significant, chronic morbidity and mortality especially in juvenile fish and is often associated with factors or stressors that further predispose fish to infection. To have 65% of this group infected is a higher-than-normal incidence of infection especially since this type of infection is underestimated in preserved fish. This constituted a significant and disturbing finding that Bay-Delta juvenile striped bass were suffering from an abnormal incidence of parasitic infections and, as such, under significant physiological stress.

Otoliths – Habitat Use

Habitat use has been shown to be an important factor in the bioaccumulation of contaminants in striped bass. This study examined habitat use in relation to the bioaccumulation and maternal transfer of xenobiotics to progeny in the San Francisco Estuary system as well as examined adult habitat patterns to better understand factors relating to the POD. Otolith microchemistry has also been used to measure habitat utilization and migration patterns of striped bass on the East Coast in the Chesapeake Bay (Secor and Dean 1992) and Hudson River (Secor and Piccoli 1996).

For this study, age and microgeochemical analyses were conducted with sagittal otoliths collected from striped bass in the Sacramento, San Joaquin, Mokelumne, American, and Feather rivers, the confluence of the Sacramento San Joaquin Rivers, Montezuma Slough, the Pacific Ocean, and San Francisco Bay. Habitat use and residence time over the life of adult striped bass were studied by analyzing strontium (Sr) isotopes using laser ablation multicollector ICP mass spectrometer (MC-ICP/MS). The last two years of female striped bass' life was examined more intensively as female striped bass transfer their entire body burden of lipophilic contaminants to the eggs at spawning. Otolith $^{87}Sr/^{86}Sr$ values indicated that the vast majority of these striped bass were residing primarily in meso-haline waters and not the ocean, which is in agreement with salinity/habitat use data.

Instantaneous growth rates determined for juvenile striped bass from all Delta regions were similar in both sampling years (2006 and 2007) and within the lower end of the range found in striped bass from Chesapeake Bay (Secor and Houde 1995). Although 2006 was considered a wet/high-flow year, and historically striped bass are found to be in better condition and in higher numbers during this type of climactic condition, the growth rates from all 2007, a very dry year, were similar. Factors affecting juvenile striped bass growth in Chesapeake Bay after ~25 days post hatching was limited by temperature and zooplankton abundance. Although, temperatures were similar in both studies (ranging from 19°C to 20.5°C in the San Francisco Estuary) for the same developmental period, primary productivity in the Bay-Delta has historically been lower than that of the Chesapeake Bay. Thus, it does not appear that this age cohort of fish from two years of investigations in the Delta as a whole were growing significantly faster or slower than their counterparts in Chesapeake Bay.

Otolith data determined that resident subpopulations of striped bass exist in some areas of the Bay-Delta. These residents would have exposures to contaminants that reflect the local habitat rather and would bioaccumulate contaminants different from migratory striped bass. On a localized scale, growth rates in a striped bass Montezuma Slough were lower in 2006 than in 2007, and the reverse was true for the Sacramento River samples where growth rates were higher in 2006 than in 2007. It is likely that juvenile striped bass are also found in different abundance and in different habitats in dry versus wet years, with some being more suitable for better growth and survival.

Conclusion

A broad variety of data associated with contaminant exposures and effects to striped bass in the Sacramento-San Joaquin River Delta and San Francisco Bay Estuary was collected to understand the life history and impacts to this species and to gain insight into the pelagic organism decline. Biomarkers and egg-chemical analyses indicated that fish were exposed to elevated chemical concentrations. Maternal transfer of xenobiotics and severe adverse effects on larval development and subsequent survival have been documented over an extended period of time, 20 years, and are likely causing population-level effects. Immuno-incompetence and morbidity compromised the health of juvenile striped bass during multiple years of investigations conducted during 1999–2007. As such, contaminant effects need to be considered a significant stressor affecting the decline of striped bass and are likely causing population-level effects in the early life stages. It must be noted that contaminant effects are one of many stressors affecting this and the other pelagic fish species in the San Francisco Estuary. This study demonstrated the variability in exposures and responses that can occur within an ecosystem and how information about the species of concern, life history, development, habitat, and chemicals are all helpful for understanding and interpreting adverse effects on biota in the environment due to multiple stressors.

Acknowledgments

Striped bass sampling and chemical analyses for this study (Ostrach 2009) was conducted by Ostrach lab staff in collaboration with the California Department of Fish and Wildlife and University of California Davis.

References

Ackermann, G.E., Schwaiger, J., Negele, R.D., and Fent, K. (2002). Effects of long-term nonylphenol exposure on gonadal development and biomarkers of estrogenicity in juvenile rainbow trout *Oncorhynchus mykiss*. *Aquat. Toxicol.* 60 (3–4): 203–221.

Arsenault, J.T., Fairchild, W.L., MacLatchy, D.L. et al. (2004). Effects of water-borne 4-nonylphenol and 17β-estradiol exposures during parr-smolt transformation on growth and plasma IGF-I of Atlantic salmon (*Salmo salar* L.). *Aquat. Toxicol.* 66 (3): 255–265.

Arzuaga, X., Wassenberg, D., Di Giulio, R. et al. (2006). The chlorinated AHR ligand 3,3′,4,4′,5-pentachlorobiphenyl (PCB126) promotes reactive oxygen species (ROS) production during embryonic development in the killifish (*Fundulus heteroclitus*). *Aquat. Toxicol.* 76 (1): 13–23.

Bailey, H.C., Ostrach, D.J. et al. (1991). Toxicity of agricultural return waters to early life stages of striped bass (*Morone saxatilis*) and Opossum Shrimp (*Neomysis mercedis*). *Proceedings from the 12th Annual Meeting of the Society of Environmental Toxicology and Chemistry (SETAC)*, pp. 50.

Bailey, H.C., Alexander, C., DiGiorgio, C. et al. (1993). The effect of agricultural discharge on striped bass (*Morone saxatilis*) in California Sacramento San Joaquin drainage. *Ecotoxicol. Environ. Saf.*

Bailey, H.C., Alexander, C., DiGiorgio, C. et al. (1994). The effect of agricultural discharge on striped bass (*Morone saxatilis*) in California Sacramento San Joaquin drainage. *Ecotoxicology* 3 (2): 123–142.

Baxter, R., Breuer, R., Brown, L. et al. (2008). Pelagic organism decline progress report: 2007 synthesis of results. Prepared by the Interagency Ecological Program for the San Francisco Estuary. https://www.waterboards.ca.gov/waterrights/water_issues/programs/bay_delta/pelagic_organism/docs/pod_ieppodmt_2007synthesis_011508.pdf.

Bennett, W.A., Ostrach, D.J., and Hinton, D.E. (1995). Larval striped bass condition in a drought-stricken estuary: evaluating pelagic food-web limitation. *Ecol. Appl.* 5 (3): 680–692.

Billiard, S.M., Timme-Laragy, A.R., Wassenberg, D.M. et al. (2006). The role of the aryl hydrocarbon receptor pathway in mediating synergistic developmental toxicity of polycyclic aromatic hydrocarbons to zebrafish. *Toxicol. Sci.* 92 (2): 526–536.

Bozcaarmutlu, A. and Arinc, E. (2004). Inhibitory effects of divalent metal ions on liver microsomal 7-ethoxyresorufin O-deethylase (EROD) activity of leaping mullet. *Mar. Environ. Res.* 58 (2–5): 521–524.

Cai, L., Klein, J.B., and Kang, Y.J. (2000). Metallothionein inhibits peroxynitrite-induced DNA and lipoprotein damage. *J. Biol. Chem.* 275 (50): 38957–38960.

Central Valley Regional Water Quality Control Board (2010). Amendments to the water quality control plan for the Sacramento River and San Joaquin River basins for the control of methylmercury and total mercury in the Sacramento-San Joaquin River delta estuary (attachment 1 to resolution No. R5-2010-0043). https://www.waterboards.ca.gov/centralvalley/water_issues/tmdl/central_valley_projects/delta_hg/2011_1020_deltahg_bpa.pdf.

Chung, M.J., Walker, P.A., Brown, R.W., and Hogstrand, C. (2005). Zinc-mediated gene expression offers protection against H_2O_2-induced cytotoxicity. *Toxicol. Appl. Pharmacol.* 205 (3): 225–236.

Cionna, C., Maradonna, F., Olivotto, I. et al. (2006). Effects of nonylphenol on juveniles and adults in the grey mullet, *Liza aurata*. *Reprod. Toxicol.* 22 (3): 449–454.

Davis, J.A., May, M.D., Ichikawa, G., and Crane, D. (2000). *Contaminant Concentrations in Fish from the Sacramento-San Joaquin Delta and Lower San Joaquin River, 1998*. Richmond, CA: San Francisco Estuary Institute.

De Parsia, M.D., Orlando, J.L., McWayne, M.M., and M.L. Hladik (2018). Pesticide inputs to the Sacramento–San Joaquin Delta, 2015–2016: results from the delta regional monitoring program. Data Series 1089. Prepared in cooperation with the Delta Regional Monitoring Program. https://pubs.usgs.gov/publication/ds1089.

Ellman, G.L., Courtney, K.D., Andres, V. Jr., and Featherstone, R.M. (1961). A new and rapid colorimetric determination of acetylcholinesterase activity. *Biochem. Pharmacol.* 7: 88–95.

Fenske, M., Maack, G., Schäfers, C., and Segner, H. (2005). An environmentally relevant concentration of estrogen induces arrest of male gonad development in zebrafish, *Danio rerio*. *Environ. Toxicol. Chem.* 24 (5): 1088–1098.

Ferreira, M., Antunes, P., and Gil, O. (2004). Organochlorine contaminants in flounder (*Platichthys flesus*) and mullet (*Mugil cephalus*) from Douro estuary, and their use as sentinel species for environmental monitoring. *Aquat. Toxicol.* 69 (4): 347–357.

Fulton, M.H. and Key, P.B. (2001). Acetylcholinesterase inhibition in estuarine fish and invertebrates as an indicator of organophosphorus insecticide exposure and effects. *Environ. Toxicol. Chem.* 20: 37–45.

George, S., Gubbins, M., MacIntosh, A. et al. (2004). A comparison of pollutant biomarker responses with transcriptional responses in European flounders (*Platicthys flesus*) subjected to estuarine pollution. *Mar. Environ. Res.* 58 (2–5): 571–575.

Guiney, P.D., Smolowitz, R.M., Peterson, R.E., and Stegeman, J.J. (1997). Correlation of 2,3,7,8-tetrachlorodibenzo-pdioxin induction of cytochrome P4501A in vascular endothelium with toxicity in early life stages of lake trout. *Toxicol. Appl. Pharmacol.* 143 (2): 256–273.

Harris, C.A., Santos, E.M., Janbakhsh, A. et al. (2001). Nonylphenol affects gonadotropin levels in the pituitary gland and plasma of female rainbow trout. *Environ. Sci. Technol.* 35 (14): 2909–2916.

Hegelund, T., Ottosson, K., Rådinger, M. et al. (2004). Effects of the antifungal imidazole ketoconazole on CYP1A and CYP3A in rainbow trout and killifish. *Environ. Toxicol. Chem.* 23 (5): 1326–1334.

Heppell, S.A., Denslow, N.D., Folmar, L.C., and Sullivan, C.V. (1995). Universal assay of vitellogenin as a biomarker for environmental estrogens. *Environ. Health Perspect.* 103 (Suppl 7): 9–15.

Hill, R.L. Jr. and Janz, D.M. (2003). Developmental estrogenic exposure in zebrafish (*Danio rerio*): I. Effects on sex ratio and breeding success. *Aquat. Toxicol.* 63 (4): 417–429.

Hodson, P.V., Efler, S., Wilson, J.Y. et al. (1996). Measuring the potency of pulp mill effluents for induction of hepatic mixed-function oxygenase activity in fish. *J. Toxicol. Environ. Health* 49 (1): 83–110.

Incardona, J.P., Carls, M.G., Teraoka, H. et al. (2005). Aryl hydrocarbon receptor-independent toxicity of weathered crude oil during fish development. *Environ. Health Perspect.* 113 (12): 1755–1762.

Ishibashi, H., Matsumura, N., Hirano, M. et al. (2004). Effects of triclosan on the early life stages and reproduction of medaka *Oryzias latipes* and induction of hepatic vitellogenin. *Aquat. Toxicol.* 67 (2): 167–179.

Ishibashi, H., Watanabe, N., Matsumura, N. et al. (2005). Toxicity to early life stages and an estrogenic effect of a bisphenol A metabolite, 4-methyl-2,4-bis(4-hydroxy-phenyl)pent-1-ene on the medaka (*Oryzias latipes*). *Life Sci.* 77 (21): 2643–2655.

Jobling, S., Coey, S., Whitmore, J.G. et al. (2002). Wild intersex roach (*Rutilus rutilus*) have reduced fertility. *Biol. Reprod.* 67 (2): 515–524.

Jobling, S., Williams, R., Johnson, A. et al. (2006). Predicted exposures to steroid estrogens in U.K. Rivers correlate with widespread sexual disruption in wild fish populations. *Environ. Health Perspect.* 114 (Suppl 1): 32–39.

Kang, I.J., Yokota, H., Oshima, Y. et al. (2003). Effects of 4-nonylphenol on reproduction of Japanese medaka, *Oryzias latipes*. *Environ. Toxicol. Chem.* 22 (10): 2438–2445.

Kester, M.H., Bulduk, S., van Toor, H. et al. (2002). Potent inhibition of estrogen sulfotransferase by hydroxylated metabolites of polyhalogenated aromatic hydrocarbons reveals alternative mechanism for estrogenic activity of endocrine disrupters. *J. Clin. Endocrinol. Metab.* 87 (3): 1142–1150.

Kidd, K.A., Blanchfield, P.J., Mills, K.H. et al. (2007). Collapse of a fish population after exposure to a synthetic estrogen. *Proc. Natl. Acad. Sci. USA* 104 (21): 8897–8901.

Kirby, M.F., Smith, A.J., Rooke, J. et al. (2007). Ethoxyresorufin-O-deethylase (EROD) and vitellogenin (VTG) in flounder (*Platichthys flesus*): system interaction, crosstalk and implications for monitoring. *Aquat. Toxicol.* 81 (3): 233–244.

Krüner, G. and Westernhagen, H.V. (1999). Sources of measurement error in assays of EROD activity of fish for biological effects monitoring. *Helgol. Mar. Res.* 53: 250–256.

Kuiper, R.V., Bergman, Å., Vos, J.G., and van den Berg, M. (2004). Some polybrominated diphenyl ether (PBDE) flame retardants with wide environmental distribution inhibit TCDD-induced EROD activity in primary cultured carp (*Cyprinus carpio*) hepatocytes. *Aquat. Toxicol.* 68 (2): 129–139.

Lacorn, M., Lahrssen, A., Rotzoll, N. et al. (2001). Quantification of metallothionein isoforms in fish liver and its implications for biomonitoring. *Environ. Toxicol. Chem.* 20 (1): 140–145.

Lionetto, M., Caricato, R., Giordano, M., and Schettino, T. (2005). Acetylcholinesterase as biomarker in environmental biomonitoring. In: *Recent Trends in the Acetylcholinesterase System* (ed. M. Parveen and S. Kumar). Amsterdam: IOS Press.

Mayer, F.L., Versteeg, D.J., McKee, M.J. et al. (1992). Physiological and nonspecific biomarkers. In: *Biomarkers: Biochemical, Physiological, and Histological Markers of Anthropogenic Stress* (ed. R.J. Huggett, R.A. Kimerle, P.M. Mehrle Jr., and H.L. Bergma), 85. Boca Raton, FL: Lewis Publishers. https://pubs.usgs.gov/publication/5210667.

McArdle, M., Elskus, A., McElroy, A. et al. (2000). Estrogenic and CYP1A response of mummichogs and sunshine bass to sewage effluent. *Mar. Environ. Res.* 50 (1–5): 175–179.

McCormick, S.D., O'Dea, M.F., Moeckel, A.M. et al. (2005). Endocrine disruption of parr-smolt transformation and seawater tolerance of Atlantic salmon by 4-nonylphenol and 17β-estradiol. *Gen. Comp. Endocrinol.* 142 (3): 280–288.

Meyer, J.N., Nacci, D.E., and Di Giulio, R.T. (2002). Cytochrome P4501A (CYP1A) in killifish (*Fundulus heteroclitus*): heritability of altered expression and relationship to survival in contaminated sediments. *Toxicol. Sci.* 68 (1): 69–81.

Miller, K.A., Addison, R.F., and Bandiera, S.M. (2004). Hepatic CYP1A levels and EROD activity in English sole: biomonitoring of marine contaminants in Vancouver Harbour. *Mar. Environ. Res.* 57 (1–2): 37–54.

Mortensen, A.S. and Arukwe, A. (2007). Modulation of xenobiotic biotransformation system and hormonal responses in Atlantic salmon (*Salmo salar*) after exposure to tributyltin (TBT). *Comp. Biochem. Physiol. C Toxicol. Pharmacol.* 145 (3): 431–441.

Mortensen, A.S., Tolfsen, C.C., and Arukwe, A. (2006). Gene expression patterns in estrogen (nonylphenol) and aryl hydrocarbon receptor agonists (PCB-77) interaction using rainbow trout (*Oncorhynchus mykiss*) primary hepatocyte culture. *J. Toxicol. Environ. Health A* 69 (1–2): 1–19.

Navas, J.M. and Segner, H. (2000). Antiestrogenicity of beta-naphthoflavone and PAHs in cultured rainbow trout hepatocytes: evidence for a role of the arylhydrocarbon receptor. *Aquat. Toxicol.* 51 (1): 79–92.

Okihiro, M.S., Ostrach, D.J., and Hinton, D.E. (1990). Histologic evaluation of larval striped bass exposed to the riceherbicides Ordram (molinate) and Bolero (thiobencarb), final report on research to California Department of Fish and Game Environmental Services Branch, Pesticide Investigations Unit: 20p + Appendices.

Ostrach, D.J. (2006). Implications of sublethal contaminant exposure on the early life stages of striped bass (*Morone saxatilis*) in the San Francisco estuary. University of California, p. 71.

Ostrach, D.J. (2009). The role of contaminants, within the context of multiple stressors, in the collapse of the striped bass population in the San Francisco estuary and its watershed. Year 2 final report prepared for the Department of Water Resources. Agreement No. 4600004664. https://www.waterboards.ca.gov/waterrights/water_issues/programs/bay_delta/docs/cmnt091412/sldmwa/ostrach_et_al_2009.pdf.

Ostrach, D.J., Low, J.M., Eder, K.J. et al. (2008). Maternal transfer of xenobiotics and effects on larval striped bass in the San Francisco estuary. *Proc. Natl. Acad. Sci.* 105 (49): 19353–11935.

Sadzikowski, M.R. and Wallace, D.C. (1974). The incidence of *Lironeca ovalis* (say)(crustacea, isopoda) and its effects on the growth of white perch, *Morone americana* (Gmelin), in the Delaware River near Artificial Island. *Chesap. Sci.* 15 (3): 163–165.

Sakamoto, T. and McCormick, S.D. (2006). Prolactin and growth hormone in fish osmoregulation. *Gen. Comp. Endocrinol.* 147 (1): 24–30.

Sandahl, J.F., Baldwin, D.H., Jenkins, J.J., and Scholz, N.L. (2005). Comparative thresholds for acetylcholinesterase inhibition and behavioral impairment in coho salmon exposed to chlorpyrifos. *Environ. Toxicol. Chem.* 24 (1): 136–145. https://pubmed.ncbi.nlm.nih.gov/15683177/.

Sarkar, A., Ray, D., Shrivastava, A.N., and Sarker, S. (2006). Molecular Biomarkers: their significance and application in marine pollution monitoring. *Ecotoxicology* 15 (4): 333–340.

Schlenk, D., Zhang, Y.S., and Nix, J. (1995). Expression of hepatic metallothionein messenger RNA in feral and caged fish species correlates with muscle mercury levels. *Ecotoxicol. Environ. Saf.* 31 (3): 282–286.

Secor, D.H. and Dean, J.M. (1992). Comparison of otolith-based back-calculation methods to determine individual growth histories of larval striped bass, *Morone-saxatilis*. *Can. J. Fish. Aquat. Sci.* 49 (7): 1439–1454.

Secor, D.H. and Houde, E.D. (1995). temperature effects on the timing of striped bass egg production, larval viability, and recruitment potential in the Patuxent River (Chesapeake Bay). *Estuaries* 18 (3): 527–544.

Secor, D.H. and Piccoli, P.M. (1996). Age- and sex-dependent migrations of striped bass in the Hudson River as determined by chemical microanalysis of otoliths. *Estuaries* 19 (4): 778–793.

da Silva, M.E. and Meirelles, N.C. (2004). Interaction of non-ionic surfactants with hepatic CYP in *Prochilodus scrofa*. *Toxicol. in vitro* 18 (6): 859–867.

Sommer, T., Armor, C., Baxter, R. et al. (2011). The collapse of pelagic fishes in the upper San Francisco estuary. *Fisheries* 32 (6): 270–277. https://pubs.usgs.gov/publication/70030175.

Spearow, J.L., Doemeny, P., Sera, R. et al. (1999). Genetic variation in susceptibility to endocrine disruption by estrogen in mice. *Science* 285 (5431): 1259–1261.

Springman, K.R., Kurath, G., Anderson, J.J., and Emlen, J.M. (2005). Contaminants as viral co-factors: assessing indirect population effects with the vitality model. *Aquat. Toxicol.* 71: 13–23.

Thorpe, K.L., Hutchinson, T.H., Hetheridge, M.J. et al. (2001). Assessing the biological potency of binary mixtures of environmental estrogens using vitellogenin induction in juvenile rainbow trout (*Oncorhynchus mykiss*). *Environ. Sci. Technol.* 35 (12): 2476–2481.

Vaccaro, E., Meucci, V., Intorre, L. et al. (2005). Effects of 17β-estradiol, 4-nonylphenol and PCB 126 on the estrogenic activity and phase 1 and 2 biotransformation enzymes in male sea bass (*Dicentrarchus labrax*). *Aquat. Toxicol.* 75 (4): 293–305.

Van Aerle, R., Pounds, N., Hutchinson, T.H. et al. (2002). Window of sensitivity for the estrogenic effects of ethinylestradiol in early life-stages of fathead minnow, *Pimephales promelas*. *Ecotoxicology* 11 (6): 423–434.

Van Campenhout, K., Bervoets, L., and Blust, R. (2003). Metallothionein concentrations in natural populations of gudgeon (*Gobio gobio*): relationship with metal concentrations in tissues and environment. *Environ. Toxicol. Chem.* 22 (7): 1548–1555.

Viarengo, A., Burlando, B., Cavaletto, M. et al. (1999). Role of metallothionein against oxidative stress in the mussel (*Mytilus galloprovincialis*). *Am. J. Phys.* 277 (6 Pt 2): R1612–R1619.

Voelker, D., Vess, C., Tillmann, M. et al. (2007). Differential gene expression as a toxicant-sensitive endpoint in zebrafish embryos and larvae. *Aquat. Toxicol.* 81 (4): 355–364.

Wassenberg, D.M. and Di Giulio, R.T. (2004). Synergistic embryotoxicity of polycyclic aromatic hydrocarbon aryl hydrocarbon receptor agonists with cytochrome P4501A inhibitors in *Fundulus heteroclitus*. *Environ. Health Perspect.* 112 (17): 1658–1664.

Wassenberg, D.M., Nerlinger, A.L., Battle, L.P., and Di Giulio, R.T. (2005). Effects of the polycyclic aromatic hydrocarbon heterocycles, carbazole and dibenzothiophene, on *in vivo* and *in vitro* CYP1A activity and polycyclic aromatic hydrocarbon-derived embryonic deformities. *Environ. Toxicol. Chem.* 24 (10): 2526–2532.

Whyte, J.J., Jung, R.E., Schmitt, C.J., and Tillitt, D.E. (2000). Ethoxyresorufin-O-deethylase (EROD) activity in fish as a biomarker of chemical exposure. *Crit. Rev. Toxicol.* 30 (4): 347–570.

Zinkl, J.G., Lockhart, W.L., Kenny, S.A., and Ward, F.J. (1991). The effects of cholinesterase-inhibiting insecticides on fish. In: *Cholinesterase-Inhibiting Insecticides: Their Impact on Wildlife and the Environment*, vol. 2 (ed. P. Mineau), 233–254. Amsterdam: Elsevier.

7

Management of Environmental Systems Using Cumulative Risk Assessment – The Importance of Multiple Stressor Cumulative Ecological Risk Assessment for Long-Term Management of Environmental Systems Demonstrated Using the Upper San Francisco Estuary Case Study

Wayne G. Landis

College of the Environment, Institute of Environmental Toxicology and Chemistry, Western Washington University, Bellingham, WA 98229, USA

Introduction

In the 1990s, Landis and colleagues had demonstrated that ecological risk assessment incorporating both multiple stressors and endpoints is possible (Landis and Wiegers 1997; Weigers et al. 1998; Landis 2003). The basic method was named the relative risk model (RRM) (Landis 2005). Since 2013, Bayesian networks (BNs) have been used to perform the calculations (Ayre and Landis 2012; Landis 2022). Duncan (2005) described the importance of connecting landscape-level endpoints with management decisions using the RRM. Burton and Rohr (2025) emphasized the importance of cumulative risk assessments in the understanding of the effects of chemicals on ecological structures. The goal of this chapter is to present an example of multiple stressor approach and to discuss the expansion of the use of multistressor or cumulative risk assessment. I am using a recent ecological risk assessment program report for the California Delta case study (Landis et al. 2024a) available at the Western Washington University CEDAR site as source material. The goal of this rendition is to present this example of multiple stressor approach and to discuss the expansion of the use of multistressor or cumulative risk assessment.

Basics of Using the RRM for Multiple Stressor Risk Assessment

The fundamental structure is based on the source-stressor-habitat (location)-effect-impact (risk) model as detailed in Landis and Wiegers (2005) as developed from (Landis and Wiegers 1997). In order to solve the issues of combining variables with different units, ranks were used that corresponded with none, low, medium, and high levels corresponding to the sources, habitat, and impacts based on specific criteria for each input. The connections between the variables were done by the use of filters (0,1) to signify whether a connection (as in exposure or effect) existed. The impact was the total score from the scores of all the pathways across all the sources, exposures, and effects. In addition to the use of ranks, the study area was broken into geographic regions with similar characteristics so that the spatial distribution of the risk could be assessed. Hart Hayes and Landis (2004) incorporated the use of Monte Carlo methods to generate probability distributions and to build a definitive risk assessment for the Pacific herring stock at Cherry Point, Washington. This method was applied until the introduction of BNs to better describe probability, uncertainty, and the interactions among the variety of inputs used to describe risk (Ayre and Landis 2012). This has become the standard approach to describe toxicity and risk using BNs ever since (Landis 2022). The ability to use ranking to describe exposures and effects the RRM has been able to incorporate exposure–response curves to describe toxicity instead of point estimates such as EC50s. LOECs, MATCs, or other summary estimates-projections. The ranking approach also makes estimating the toxicity of mixtures straightforward, as was done in the beginning with mixtures of polycyclic aromatic hydrocarbons (PAHs) based on measured concentrations in the Port of Valdez (Wiegers et al. 1998).

In contrast, conventional risk assessments at this time that used point estimates to describe toxicity were not spatially explicit and did not incorporate multiple stressors. This contrast will be addressed in more detail in the discussion section of this chapter.

The Use of BNs for Calculation

Multiple contaminants can be found in the region along with issues of temperature and other water quality and water availability issues. The purpose of this study is to demonstrate that mixture toxicity can be estimated, and risk calculated using a BN relative risk model (BN-RRM). Exposure–response models have been used to describe toxicity in BNs since Landis et al. (2017a, 2020) incorporated the equations and the derived conditional probability tables (CPTs) as part of the process. Since 2013, we have published a series of papers using the BN-RRM (Landis 2021) and BNs have become widely used in environmental assessments (Kaikkonen et al. 2021).

Since the mid-2000s, BNs have been applied to environmental management, risk assessment, and research and monitoring to inform decision-making and resource management (Marcot et al. 2006; Nyberg et al. 2006; Pollino et al. 2007; Uusitalo 2007; Carriger et al. 2016). Landis (2021) and Kaikkonen et al. (2021) have reviewed the status of the practice for the BN-RRM and the broader application of BNs, respectively.

Ayre and Landis (2012) demonstrated that the causal framework of the RRM translates directly into the tiered node structure of a BN. A series of papers estimating risk due to mercury in the South River in Virginia established the applicability of the BN-RRM to estimate risk to organisms and water quality (Landis et al. 2017a), assess risk to human health and well-being (Harris et al. 2017), evaluate management alternatives, and inform adaptive management decisions (Johns et al. 2017; Landis et al. 2017b). There are specific advantages to the use of the BN-RRM for multiple stressors.

First, BNs innately incorporates categories and, as in the case of the RRM, ranks to describe systems. Second, interactions between multiple stressors can be combined using several pathways and the CPTs to calculate outcomes. Entropy analysis was the method used to document model sensitivity. As with the RRM, the method has now been applied to a wide series of sites and questions, from forestry management to invasive species, to disease, the interaction of ecological and human health endpoints, the flows of large rivers, and now the efficacy and risks of synthetic biology.

Third, with appropriate software, it is straightforward to build CPTs directly from data using a number of statistical methods and machine leaning tools. Likewise, sensitivity analysis ranking the nodes with the most influence on a specific output (endpoint) distribution is part of the software.

Fourth, exposure–response equations derived from software such as the DRC package written in R (Ritz et al. 2015) can be used to calculate effects, although mixtures of chemicals, water quality parameters, and climate change remain a challenge (Landis et al. 2023).

Fifth, the graphical output is particularly valuable in presenting the results to an audience of managers and other stakeholders. The models can be presented in real time and used to demonstrate the outcomes of different management alternatives. This is particularly useful in the case of multiple stressors. The values of the input variables, such as water temperature or mercury concentration, can be altered in real time and the changes in the risk viewed immediately. This is particularly valuable when there is contention regarding causes of a decline or the effectiveness of a management alternative.

Sixth, it is also possible to calculate input scenarios that would reduce risk to a lower level. The back calculation can provide probability distributions that would result in meeting management goals. In some cases, relatively small changes to a few inputs are required as compared to a magic bullet fix. Such a model would facilitate adaptive management (Landis et al. 2017).

In this chapter, I am using the Upper California Estuary study report as a case study to illustrate a multiple stressor risk assessment (Landis et al. 2024a). There are also specific suggestions as to endpoint selection and long-term monitory programs that would facilitate the long-term management of the environment.

The Upper San Francisco Estuary Example – The Study Site

The *Upper San Francisco Estuary (USFE)* drains the entirety of the Central Valley of California with many different contaminants ending up in Suisun Bay and the Delta. Agricultural and urban land use practices are the primary sources for these contaminants. Contaminants have long been considered a threat to fish and other aquatic organisms in the Suisun/Delta region of the USFE. The USFE contains key species and ecosystem services. The delta smelt is a key forage fish endemic to California and is only present in the San Francisco Estuary. Chinook salmon are an iconic species and many runs pass through the USFE to spawning grounds upstream. The macroinvertebrate community is a food resource to multiple fish and other species. The habitats in the region support these and numerous other bird, mammal, amphibian, and insect species and provide recreational opportunities and water for irrigation, drinking, and transportation.

Geographic Extent. The study area is located in the Central Valley of California and encompasses an area of approximately 3,441 km^2. It is delineated by the Legal Delta Boundary established under the Delta Protection Act (Section 12220 of the Water Code) (CDWR 2020a) and the Suisun Boundary, Conservation Zone 11, as defined by the Bay Delta Conservation Plan. In total, the area includes the southern half of the Sacramento River watershed, the northern half of the San Joaquin River watershed, the Delta, Suisun Bay, Suisun Marsh, and its watershed.

Risk Regions. The study area was divided into six smaller sub-(risk) regions based on hydrological delineations and the land use similarities within a region. Boundary lines follow those delineations. The resulting risk regions, from north to south, are North Delta, Sacramento River, Central Delta, and South Delta, and from east to west are Confluence and Suisun Bay.

The **North Delta** risk region is delineated by the Legal Delta Boundary on its north and west border. Its east border includes the Sacramento Deep Water Ship Canal and is adjacent to the western border of the Sacramento River risk region. The risk region encompasses the southwest portion of Yolo County and the eastern portion of Solano County.

The **Sacramento** risk region is directly east and adjacent to the North Delta region, sharing its western border, the Sacramento Deep Water Ship Canal, with it. Its east border extends south along the Legal Delta Boundary and terminates at the northern boundary of the Central Delta risk region. This risk region encompasses the southeastern portion of Yolo County, the southwestern portion of Sacramento County, and the southeastern portion of Solano County.

The **North Delta** risk region is delineated by the Legal Delta Boundary on its north and west borders. Its east border includes the Sacramento Deep Water Ship Canal and is adjacent to the western border of the Sacramento River risk region. The risk region encompasses the southwest portion of Yolo County and the eastern portion of Solano County.

The **Sacramento** risk region is directly east and adjacent to the North Delta region, sharing its western border, the Sacramento Deep Water Ship Canal, with it. Its east border extends south along the Legal Delta Boundary and terminates at the northern boundary of the Central Delta risk region. This risk region encompasses the southeastern portion of Yolo County, the southwestern portion of Sacramento County, and the southeastern portion of Solano County.

The **Central Delta** risk region borders the Confluence to the west and the study site boundary to the east. Its southern boundary includes the Clifton Court Forebay, Union Island, and Robert's Island-Trapper Slough watersheds that delineate the northern border of the South Delta risk region. The Central subregion northern border is delineated by the Three-mile Slough, South Mokelumne River, and Hog Slough watershed that forms the southern border of the Sacramento risk region. The risk region encompasses the southwestern portion of Sacramento County, the northeastern portion of Contra Costa County, and the eastern portion of San Joaquin County.

The **South Delta** risk region shares its northern border with the Central Delta region, whereas its east, south, and western borders are delineated by the Legal Delta Boundary's southeastern, south, and southwestern boundaries. The risk region encompasses the southwestern portion of San Joaquin County and the northeastern portion of Alameda County.

The **Confluence** is bordered on the west by the Suisun Bay risk region on the north and south by the Legal Delta Boundary, and on the east by the Central Delta risk region. The eastern border originates in the south at the Lower Marsh Creek watershed border and extends north to the beginning of the Sacramento Deep Water Ship Canal. The region encompasses the southwestern portion of Sacramento County and the northeastern section of Contra Costa County.

The **Suisun Bay** risk region is defined on its north, south, and west borders by the Suisun Boundary. It shares its eastern border with the Confluence risk region that originates south near Shore Acres and extends northeast to the south edge of the Lucol-Hollow watershed near Montezuma Hills. Most of the region is in the southeastern section of Solano County with the Contra Costa County along its southern border.

The Development of the Conceptual Model

The conceptual model (CM) (Figure 7.1) is based on the sources-stressors-habitat-effects-impacts structure of the BN-relative risk model for risk assessment. The fundamental structure has proven appropriate in a number of previous studies on marine and freshwater systems including Padilla Bay, WA (Herring et al. 2015), South River, VA (Johns et al. 2017; Landis et al. 2017a), estuaries near Brisbane, Australia (Graham et al. 2019), and four rivers/estuaries in Puget Sound (Landis et al. 2020; Mitchell et al. 2021). Use of the BN-RRM structure facilitates the assessment of the available data sources to build quantitative cause–effect models in order to estimate risk. The BN used to do the calculation is derived directly from this CM.

Building of the CM starts at each end. The Sources describe the entities in the study area that generate the stressors of interest in the risk assessment.

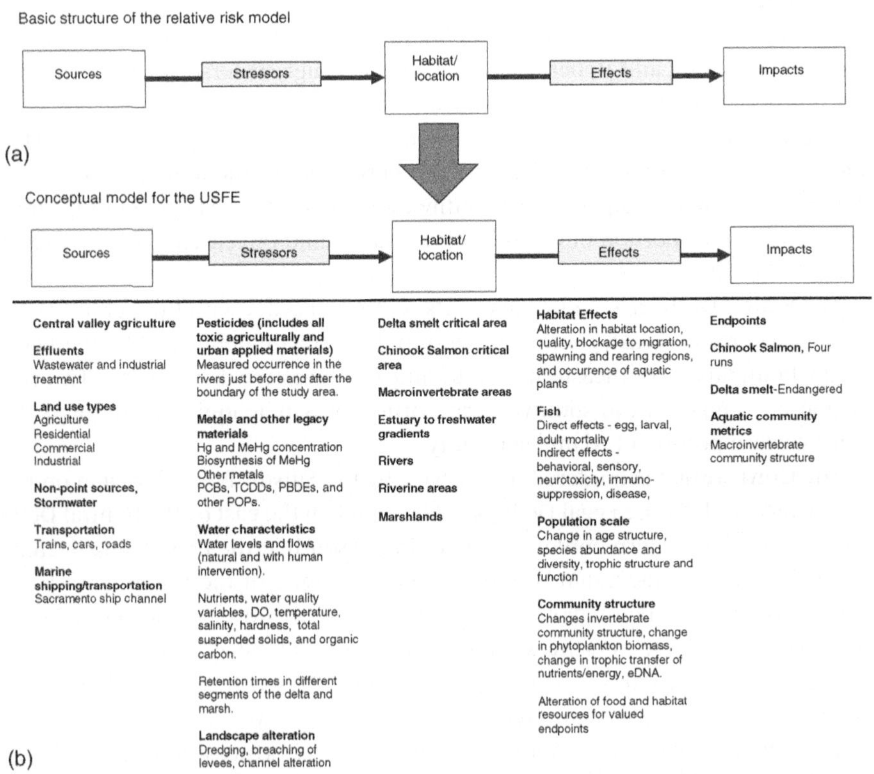

Figure 7.1 Conceptual models. (a) The original relative risk model for ecological risk assessment. (b) The basic format populated with site-specific factors for the Upper San Francisco Estuary study area. The conceptual model is the basis for the derivation of the Bayesian network.

The Impacts section is a listing of the endpoints that, by definition, have an importance to the stakeholders and managers of the site. In this study, the Sources are Central Valley Agriculture, Effluents, Land Use Practices, Stormwater Runoff, Transportation, and Marine Shipping/Transportation.

Central Valley agriculture is a source of pesticides, nutrients, and other contaminants entering via the tributaries and due to activities of this key California industry. Effluents constitute the various regulated point sources in the study area. Some of the effluents are municipal wastewater from residential areas, some are industrial, and others may be a combination. Land use information is important in characterizing the types of other inputs to the system, some by nonpoint source contributions, stormwater inputs, and habitat alteration for the endpoint species. Transportation includes trains, cars, roads, and the materials that they release to the aquatic system. There is also a large amount of marine shipping in the study area, including many shipping channels and other waterways.

At the other end of the model are the endpoints chosen for this initial study. Chinook salmon are an iconic species in the region, highly managed and regulated. Using the entity–attribute system of defining an endpoint, Chinook is the entity, and the attribute is survivorship of the in- and out-migrating fish. Delta smelt are endangered and an iconic species for the USFE. The entity is the delta smelt, and the attributes are habitat quality and population abundance. Macroinvertebrate community structure was included in the initial formulation. Subsequently, in our evaluation of the data, it was found that contaminants were not drivers of the community structure, but other water quality parameters, such as temperature and turbidity, influence macroinvertebrate communities.

The Sources and Impacts bracket the cause–effect framework in the CM. In between the two, the Stressor and Habitats/Location nodes estimate the exposures from multiple stressors across the USFE. The Stressors include pesticides, metals, legacy contaminants, the water quality characteristics of the USFE, and alterations to the landscapes and waterways from dredging, breaching of levees, and other activities. The Habitat/Location section sets the spatial areas where the stressors and endpoints intersect. These areas are mapped within each risk region. Delta smelt and Chinook salmon habitats are key since they directly affect the endpoints. Macroinvertebrate habitat is also related directly to each endpoint. Other habitats include the transition zones from estuary to subbasins, to rivers, to riverine habitats, and finally to marshlands.

Effects is the final section and lists the variety of effects that are initiated by exposure to ecological stressors. First is habitat effects, as in change in location, addition due to restoration activities, blockage to migration, and other physical effects. There are also direct effects like toxicity, both acute and chronic, and indirect ones such as populations and communities.

The next step is to evaluate the sources of information available to populate the framework and to perform the risk calculations. The next sections describe the sources of information and how they apply.

Pesticide Data

The SURF datasets contain data on the presence or absence of pesticides in California's surface waters and sediments (CDPR 2019; SFEI and ASC 2019). Aqueous pesticide data from 2010 to 2020 were extracted from both databases and then sorted by the sampling site locations to identify chemical contaminants and their concentrations in each risk region. The 10-year aqueous concentration data for each chemical in each risk region was then compared to USEPA's Office of Pesticide Program's Aquatic Life Benchmarks for Freshwater Species (USEPA 2020a). The benchmarks were based on aquatic toxicity tests from scientific studies and used to assess the effects of chemical contaminants in surface waters on aquatic life.

Water Quality Data

The CEDEN and CDWR datasets contain information about California's surface waters including water quality, aquatic habitat, and wildlife data (SFEI and ASC 2019; CDWR 2020b). Similar to the SURF data, CEDEN data from 2010 to 2020 were extracted and sorted by risk region. Data included water quality measurements; chemical contaminant concentrations; species abundance data for algae, cyanobacteria, annelids, crustaceans, mollusks, and other benthic and water column organisms; and habitat data for marshlands, open channel, and bank/riparian areas.

Chinook, Delta Smelt Trawl Data

Several fish sampling and monitoring surveys take place throughout the USFE and the Delta by the U.S. Fish and Wildlife Service and the California Department of Fish and Wildlife. The scope of this program has been expanded to include surveys that specifically monitor the abundance of the delta smelt, and other species associated with pelagic organism decline (POD): Chinook, steelhead, striped bass, threadfin shad, and longfin smelt (U.S. Fish and Wildlife [USFW] 2020). These datasets were used to populate the catch per unit effort nodes in the Fish BN.

The **Kodiak Trawl** operates from January to May and samples at 40 different sites (CDFW 2020c). The Kodiak Trawls target adult spawning delta smelt. Operated since 1967, the Midwater Trawl is the longest running of the fish monitoring

Table 7.1 Chemicals considered for the Predicted Fish Mortality pathway.

Contaminant	Included	Reason	Dose-Response Data	Organism	Response Measured
Malathion (mixture)	Yes	Sufficient dose-response data (and metadata) and environmental concentration data	Laetz et al. 2009	Chinook Salmon	AChE inhibition to Mortality
Diazinon (mixture)	Yes	Sufficient dose-response data (and metadata) and environmental concentration data	Laetz et al. 2009	Chinook Salmon	AChE inhibition to Mortality
Chlorpyrifos	Yes	Sufficient dose-response data (and metadata) and environmental concentration data	Hutton et al. 2021	Inland Silverside	Mortality
Bifenthrin	Yes	Sufficient dose-response data (and metadata) and environmental concentration data	Hutton et al. 2021	Inland Silverside	Mortality
Mercury	Yes	Sufficient dose-response data (and metadata) and environmental concentration data	Dillon et al. 2010	Fathead Minnow, Brook Trout, and Rainbow Trout	Percent Injury
Myclobutanil	No	Sufficient dose-response data (and metadata) but records of environment only show 6 values over zero in the last 10 years. All these values are far below EPA aquatic life benchmarks for this contaminant. This means uncertainty in the node is driving the results to appear far higher than they are	Hutton et al. 2021	Inland Silverside	Mortality
Triadimefon	No	Sufficient dose-response data (and metadata) but records of environment show zero values over zero in the last 10 years. This means uncertainty in the node is driving the results to appear far higher than appropriate.	Hutton et al. 2021	Inland Silverside	Mortality
Dicloran	No	Sufficient dose-response data (and metadata) and but no environmental concentration data	Hutton et al. 2021	Inland Silverside	Mortality
Paraquat	No	Sufficient dose-response data (and metadata) and but no environmental concentration data	Hutton et al. 2021	Inland Silverside	Mortality
Penconazole	No	Sufficient dose-response data (and metadata) and but no environmental concentration data	Hutton et al. 2021	Inland Silverside	Mortality

trawls. Initially comprised to map the abundance and distribution of striped bass, the Midwater Trawl now provides data on the relative abundance of POD species. The USFW combines multiple trawl surveys to provide fish abundance estimates (USFW 2020). Of these surveys, three specific trawling locations that use either the Kodiak and/or the Midwater methodology take place to capture salmonid and other fish movements throughout the Delta.

Pesticide data in water samples collected from 2010 to 2020 in the study area were compiled and sorted by the sampling site locations to identify chemical contaminants and their concentrations in each risk region. The data for each chemical in each risk region was then compared to USEPA's Office of Pesticide Program's Aquatic Life Benchmarks for Freshwater Species (USEPA 2020a). Those chemicals that exceeded the benchmark concentrations to cause acute or chronic adverse effects in fish and invertebrates, or acute effects in vascular or nonvascular plants in each risk region, were then tabulated (Table 7.1).

Additional Chemical Screening and Geographic Distribution

Another means of understanding the distribution of contaminants in a region is to compare the frequency of a recognized regulatory limit and the location, or risk region, where it takes place. Three chemicals (Figure 7.2) were investigated in this risk assessment. Using these graphs, it is possible to view the samples, the number of times the samples exceeded that limit by location and month, and the statistical distribution of those occurrences via the box and whisker plots. At a glance, it is possible to detect that the Confluence has multiple exceedances for bifenthrin in it and that mercury is found in Suisun Bay. It is also easy to see the variation in number of measurements taken between the regions indicating uneven sampling of the site.

Building the Predicted Fish Mortality Pathway

A key component of the risk assessment is the predicted fish mortality pathway. This is the part of the model that estimates the mixture toxicity to the fish endpoints.

The Predicted Fish Mortality node integrates the predicted toxicity from each pathway and is the node that when combined with the CPUE (catch per unit effort) estimates probable fish mortality for each of the risk regions. A CPT was derived to integrate the three pathways and estimate the overall chemical toxicity. The frequency of each concentration was based on the 10 years of data in our combined dataset and during only the wet season. The mixture mortality for each pathway was then illustrated and the Predicted Fish Mortality was calculated.

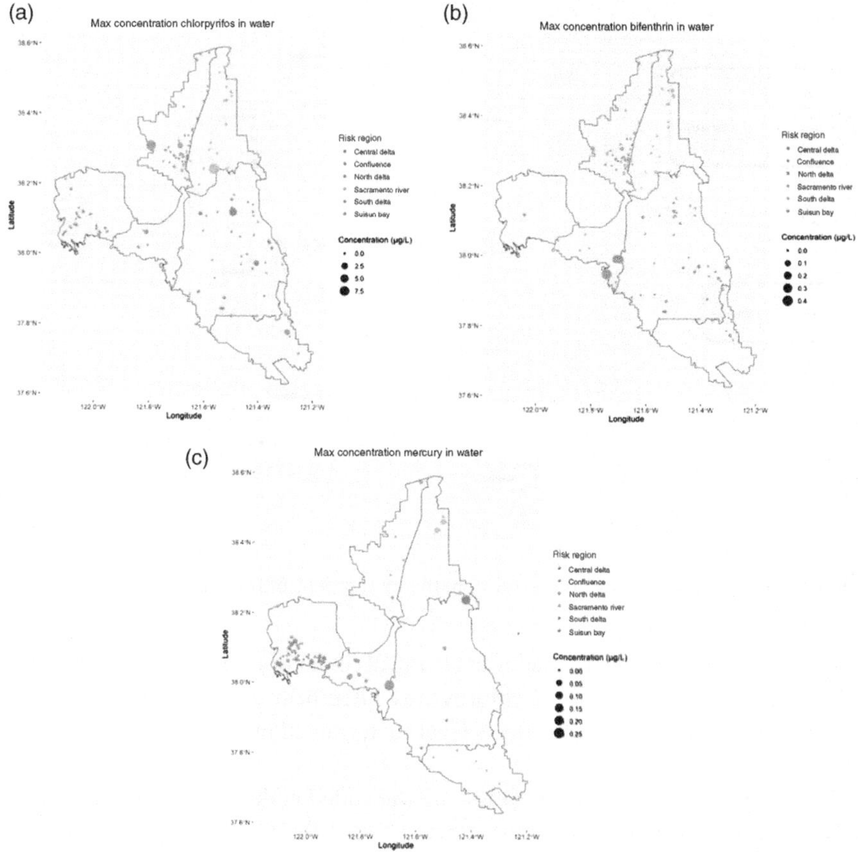

Figure 7.2 Chemical distribution and concentrations by risk region. The sample sites and the maximum concentrations for chlorpyrifos (a), bifenthrin (b), and mercury (c). Note the distribution of the sampling sites, the variability in maximum chemical concentration, and the variability among the risk regions. *Source:* Landis et al. 2024 / Western Washington University.

In the prototype BN, the fish mortality pathway consisted of malathion and diazinon that were known synergists and AChE inhibitors. Additionally, the chemicals described using the additive mixture model were chlorpyrifos, bifenthrin, myclobutanil, and triadimefon. For the final analysis, myclobutanil and triadimefon were dropped because of the lack of sufficient information.

The final Predicted Fish Mortality network (Figure 7.3) illustrates the layout of the segment of the BN. The Risk Region node allows the calculation of fish mortality by selecting the appropriate dataset from each of the six risk regions. The Season node selects the data from the wet or dry seasons at the site. Next are

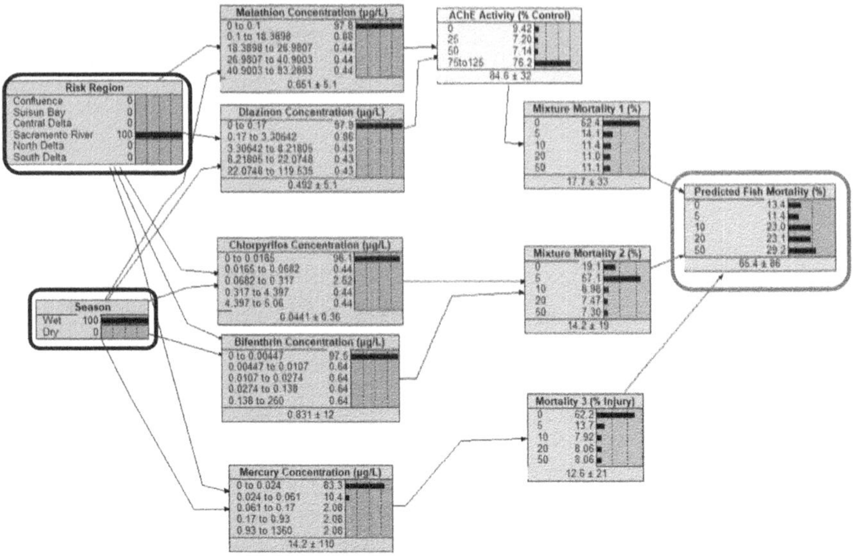

Figure 7.3 The Predicted Fish Mortality pathway segment of the Bayesian network.

the nodes describing the frequency of chemical concentrations in each of the sites. These are malathion, diazinon, chlorpyrifos, bifenthrin, and mercury. The values were discretized according to effects levels determined by the exposure–response information.

Season and Mercury Concentration are the added nodes. Myclobutanil and triadimefon have been removed. As shown, no specific Risk Region or Season has been selected. The output shown is an example from the additive CPT.

There are three different mixture pathways, namely, Mixture Mortality 1 (MM1), Mixture Mortality 2 (MM2), and Mortality 3 (MM3). MM1 describes the toxicity of malathion and diazinon, which are known synergists (Laetz et al. 2009). MM2 describes the mixture toxicity of chlorpyrifos and bifenthrin using the mixture addition model described previously. M3 describes the toxicity of mercury to fish using the dataset compiled by Dillon et al. (2010).

The nodes for Malathion Concentration, Diazinon Concentration, and so on (blue nodes) represent the frequency of each concentration for the 10-year sampling period. The listed concentrations for each are different since they are based on segments represented by the individual exposure–response curves. The segments correspond to the 0%, 5%, 10%, 20%, 50%, and over mortality rates and are combined in the appropriate Mixture Mortality node (green nodes). In the case of malathion and diazinon, the values correspond to the AChE activity (yellow node) values that in turn correspond to mortality.

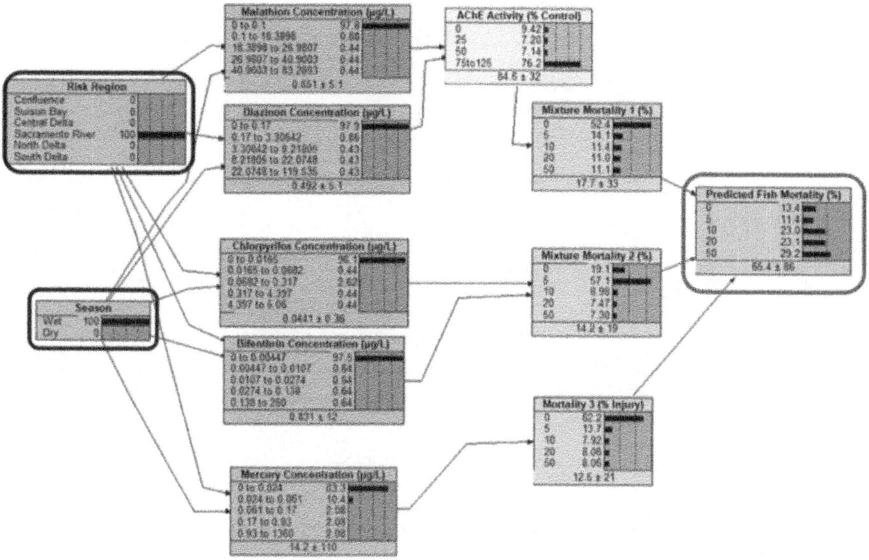

Figure 7.4 Calculation of Predicted Fish Mortality for the Sacramento River Risk Region and the wet season. The calculation is that 52.3% of the time the toxicity would equal to or exceed the EC20. The output is from the additive CPT.

It is possible to estimate the Predicted Fish Mortality for each of the risk regions. The next example illustrates the Predicted Fish Mortality for the Sacramento Risk Region during the wet season (Figure 7.4). In this manner, the Predicted Fish Mortality can be estimated for each risk region and compared.

As in the first example, a sensitivity analysis (mutual information) was performed to evaluate the sensitivity of the outcome (Predicted Fish Mortality %) to the nodes in the model. Sensitivity analysis of the Predicted Fish Mortality pathway was run assuming additivity of the three pathways (Figure 7.5).

Three regions, namely, Central Delta, Confluence, and North Delta have similar patterns in that the three mixture toxicity nodes are important in the estimation. The nodes combining the mixtures are coded in green. The Mercury mortality node is the most important for all the three. Then the malathion and diazinon mixture followed by bifenthrin and chlorpyrifos. Season was important only in the Confluence risk region.

The three regions, namely, Sacramento River, South Delta, and Suisan Bay, have different patterns. In two (Sacramento River and Suisun Bay), the malathion + diazinon pathway is the most important. In the South Delta, the three mixture nodes are close in value. For the Confluence and Suisan Bay models, the Season node is important.

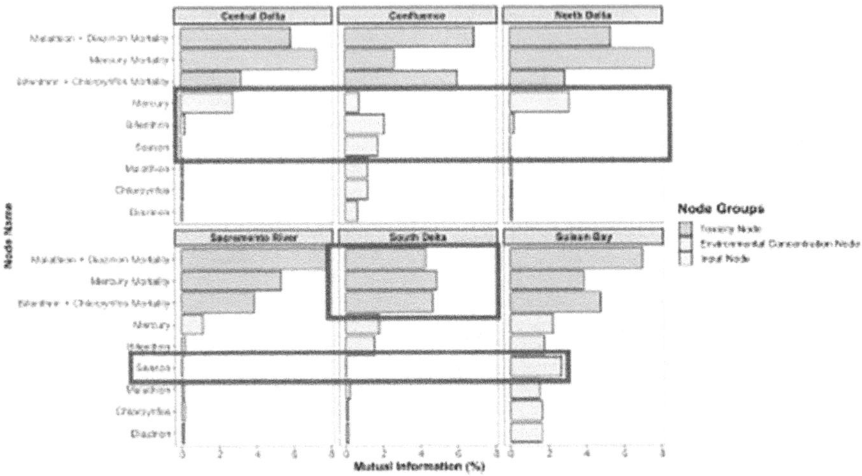

Figure 7.5 Sensitivity analysis for the prediction of the fish toxicity output using mutual information. The yellow node is in season. The blue nodes are the chemical concentrations, and the green represents the three pathways. Note that it is the mixtures that are driving the risk.

Options for the Conditional Probability Tables of the Predicted Fish Mortality Node

A specific challenge in this risk assessment has been how to combine the three different pathways to generate a prediction of mortality of fish. The examples shown so far assume that the combinations are additive, that is, MM1 + MM3 + M3 equals to the Predicted Fish Mortality. That is, if you have a value equal to an EC10 (an effect concentration equals to 10%) for all pathways, the total would be an EC30 or fitting between the EC20 and EC50 in the node. Another approach would be to take an average of each value so that the EC10 values would be averaged, and the value would be 10%. An average of an EC10 + EC20 + EC5 would be EC11 value or between the 10% and 20% value.

A third approach was to use an expert judgement method, where the calculation would be based on a CPT based on the elicitation around an exposure–response curve (see the mixtures section). The strategy for eliciting such a CPT is called "peg the corners." In other words, the corners of the table that will be the extremes of each input value would be filled and then expert judgement applied for the intermediate cases. In a typical exposure–response curve, there is little change in toxicity at lower concentrations. At a certain point, there is a steep rise in toxicity with concentration that passes through the EC50 to the EC100. Our assumption is that any value above the EC50 value is to be avoided so our effort

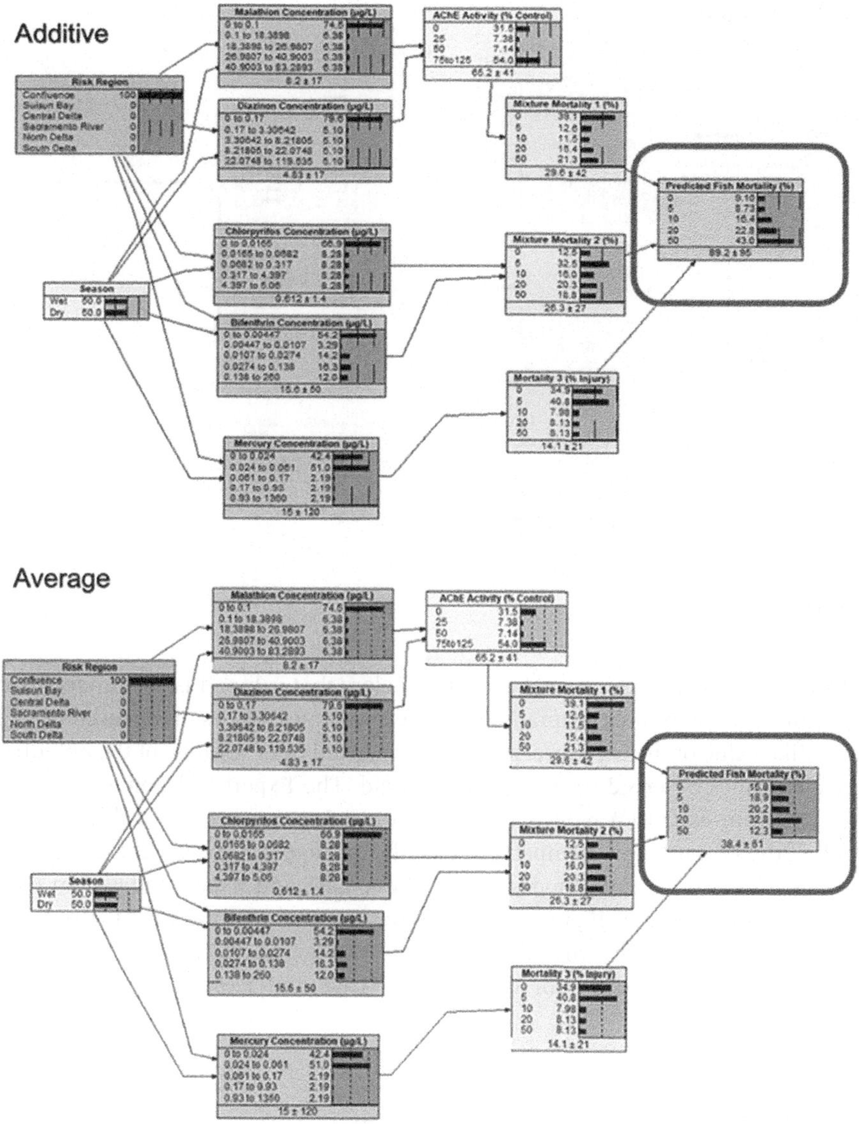

Figure 7.6 Comparison of outputs between the CPT options for the Confluence Risk Region. The Additive CPT has the highest total toxicity, the Average CPT the lowest, and the Expert Judgement the broader distribution.

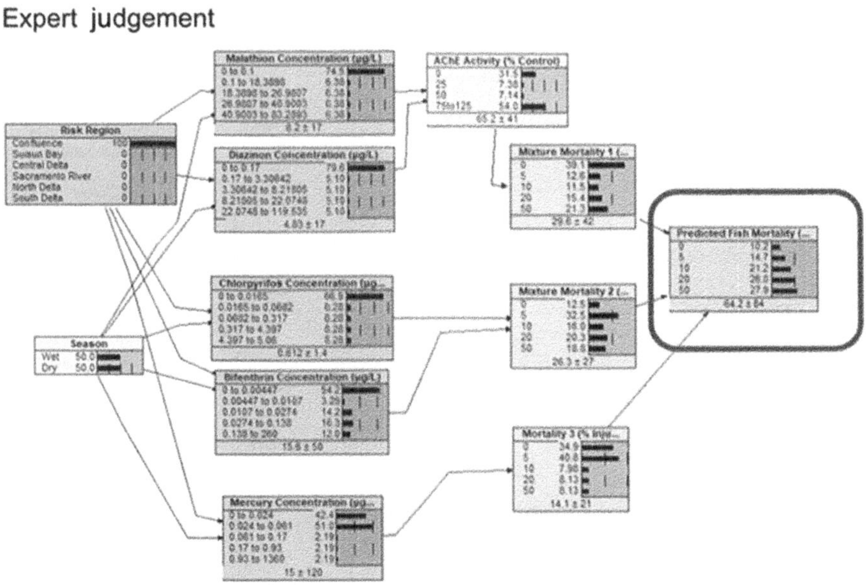

Figure 7.6 *(Continued)*

was concentrated on the part of the curve below that value. It is also understood that the regression tools generally used to estimate the values are focused on the most like value of the exposure–response curve and not the range of the potential toxicity values observed in the test or the field. The Expert Judgement CPT was derived from expert (Wayne G. Landis) knowledge was used as a third choice and all the outputs were compared. Since the average was merely the arithmetic average of the three input nodes, the Additive CPT and Expert Judgement CPT were the two choices. Figure 7.6 illustrates the calculations within the three different models and the three different outputs. Figure 7.7 shows the differences in the sensitivity analysis for the Additive CPT and Expert Judgement CPT (Figure 7.8).

Calculation of Risk to the Individual Species

The next step is to add the specific fish endpoints to the risk assessment modelThe Chinook Salmon Loss and Delta Smelt Loss nodes have been added). The numbers of fish are represented by observed CPUE for each of the species within each of the risk regions. There are a large number of species with the entry "not observed" in the catch information. It may be a better strategy to use the total number of fish caught instead.

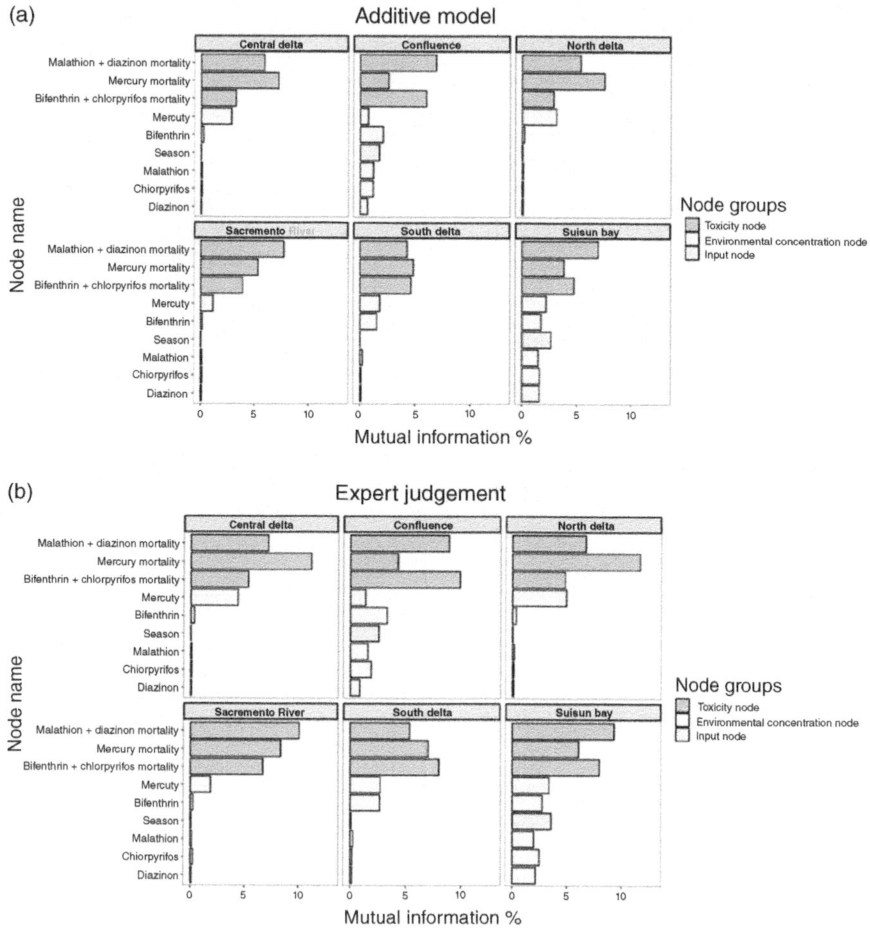

Figure 7.7 The difference in sensitivity between the Additive (a) and the Expert Judgement derived (b) models by risk region. Both models have similar sensitivities to nodes except that the Expert Judgement model had greater mutual information values in each case.

The results (Figure 7.9) do not appear to be sensitive to the Risk Region node: (a) Fish Risk model with even distribution for Risk Region, Expert Judgement CPT; the most common count for each species is the 0-to-1 category. (b) Fish Risk model Suisun Bay Risk Region, Expert Judgement CPT; the most common count for each species is the 0-to-1 category. Because of the low numbers, the change in the population numbers are in the lowest category. This is the case although the nodes are based on historic data as to the observed CPUEs for each species.

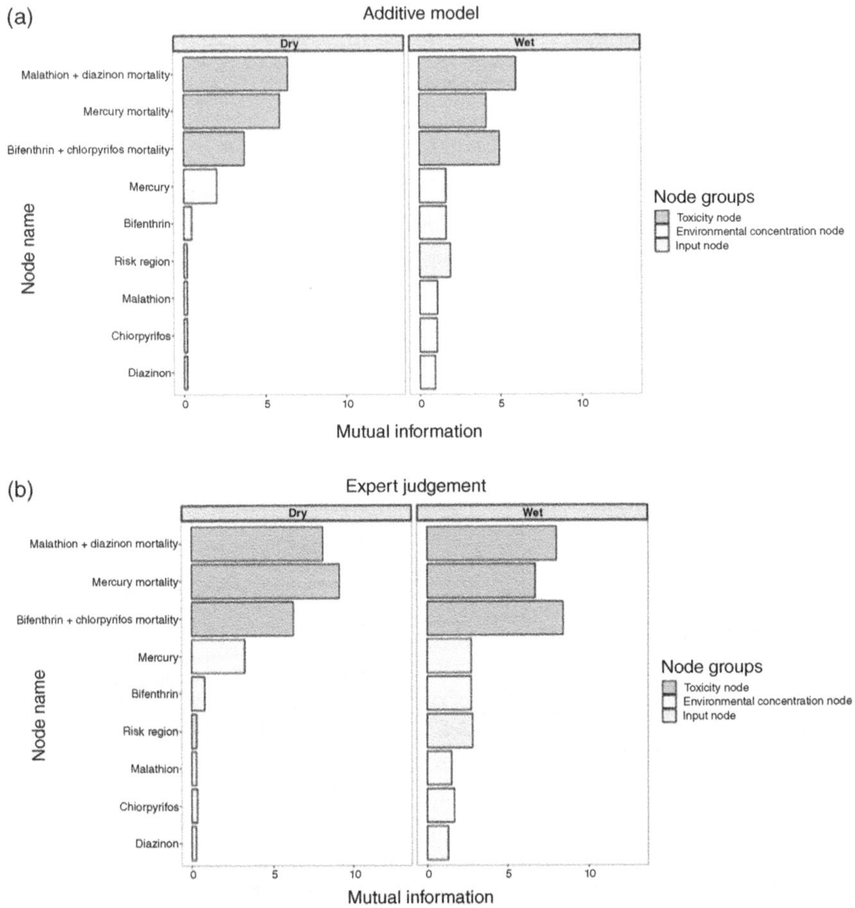

Figure 7.8 The difference in sensitivity between the Additive (a) and the Expert Judgement (b) derived models by wet or dry season. Both models have similar sensitivities to nodes except that the Expert Judgement model had greater mutual information values in each case.

Discussion: Estimating Risk with Multiple Stressors/Cumulative Risk Assessment

The previous sections summarized from Landis et al. (2024a) is one of the few examples of a multistressor risk assessment at such a scale. It does serve as one example of such an approach. Other examples exist. The end of the discussion lists advantages and drawbacks of such a design.

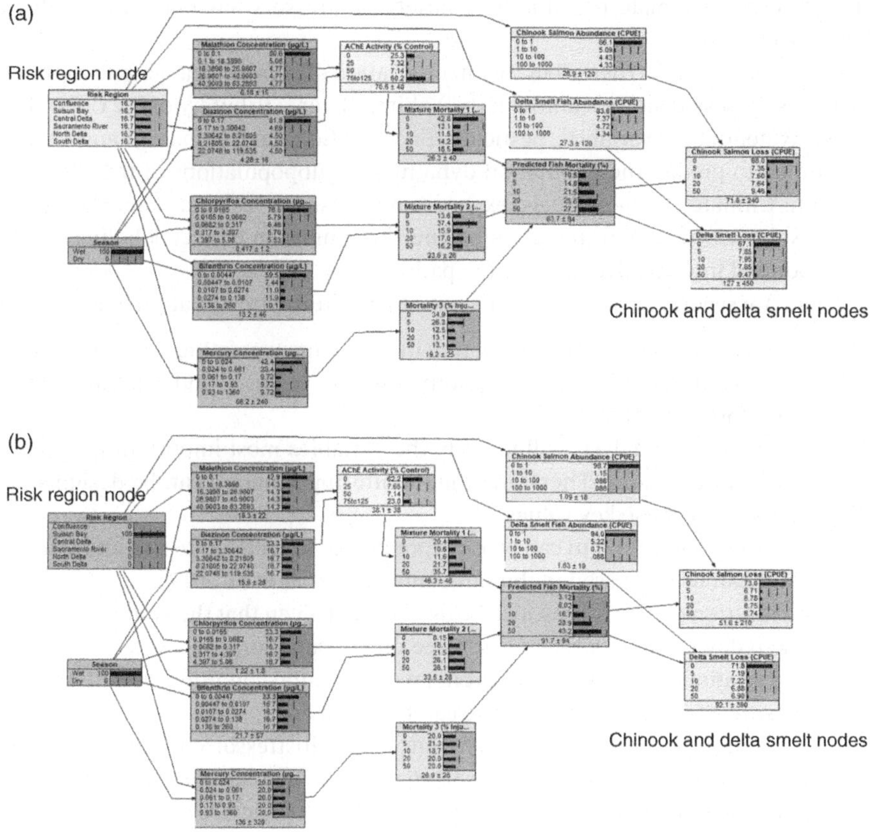

Figure 7.9 Risk calculation for each species. In (a) each of the risk regions is selected giving an overlook of the risk. In (b) the Suisan Bay risk region is selected. Note the similarity in the two distributions for each species.

Burton and Rohr (2025) have called out the need to use cumulative risk assessment to estimate risk as defined as the probability of effects to ecological endpoints valued by stakeholders. The author of this chapter concurs with this analysis. As described in this chapter, multiple stressor risk assessments have been conducted and published since the late 1990s using the RRM. The stressors have included a variety of chemicals, invasive species, disease, climate change, forest management alternatives, nutrients, microplastics, fire, insect pests, and so on. Many of the studies summarized here deal with very different types of stressors including pesticide and water temperature, parasitic disease and water quality, multiple invasive species into a site, and the population dynamics of specific gene drives effects on predator–prey dynamics in a marine reserve. The use of BNs to estimate risk due to a variety of stressors has made the analysis straightforward.

Landis et al. (2024b) describes the incorporation of climate change into a cumulative risk assessment for salmon in the Yakima River, a tributary of the Columbia River. Water quality variables, pesticide concentrations, and climate changes were combined to predict the population dynamics of subpopulations of the Chinook salmon. Climate change models, exposure response curves, and metapopulation models described salmon populations and the inhibition of acetylcholinesterase incorporated into an adverse outcome pathway.

There are several advantages to the use of a multiple stressor approach:

(1) One risk assessment can incorporate the risks incorporating the interactions among contamination, water quality, climate change, and habitat quality within the same model.
(2) The sensitivity analysis will identify the variables most important in driving the risk calculations. Therefore, the monitoring program can be designed to parameterize these key values.
(3) More than one endpoint can be incorporated so that effects to fish and human health can be addressed in the same model (Harris et al. 2017).
(4) The location of the risk can also be determined given that the system has been broken down into risk regions or some other spatial organization.
(5) The variation in risk due to different environmental conditions modifying the exposure–response of the mixture can be addressed.
(6) As Landis et al. (2017) has pointed out, the multistressor/cumulative models can form the computational basis of an adaptive management system.
(7) Epistemic uncertainty is inherently built into the structure of risk assessment model when using BNs or other probabilistic methods.

There are drawbacks as well. Part of the issue is that historic research for decision-making has been based on a single chemical-stressor approach being described by a point-derived threshold.

(1) The approach requires training in the construction of the models and in the building of BNs or the incorporation of probability distributions.
(2) Exposure–response data are not published. The literature is full of point values derived from hypothesis testing (LOECs, NOELs) or EC50 values without the entire dataset being made available for the fitting of a suitable regressions.
(3) Synergistic/antagonistic data are often lacking for different combinations of chemicals and conditions with an exposure–response model. Knowing that something demonstrates a statistically significant interaction is not sufficient to derive a quantitative function from which to model exposure–response.
(4) Field monitoring programs are not often designed to describe so many interactions over space and time. The intensity and distribution of the field sampling

needs to take into account that enough information is available to build the required probabilistic models to generate risk estimates for a wide range of inputs.

(5) Please ensure that environmental data (temperature, pH, Dissolved Oxygen, concentrations of pollutants) is take in proximity to the sampling for species occurrence and their number. The use of environmental DNA to identify a wide range of organisms would facilitate the understanding of the interactions between biota and contaminants.

The main point is that it is possible to address realistic scenarios with probabilistic models that detail interactions and can identify gaps in knowledge and technique. The mechanics of conducting multiple stressor/cumulative risk assessments are known. As in the case study with the USFE and the studies cited in this chapter, these tools can be used to gain insight into risk.

References

Ayre, K.K. and Landis, W.G. (2012). A Bayesian approach to landscape ecological risk assessment applied to the Upper Grande Ronde watershed, Oregon. *Hum. Ecol. Risk Assess.* 18: 946–970.

Burton, G.A. and Rohr, J.R. (2025). A plea for cumulative stressor risk assessments in light of climate change. *Environ. Sci. Technol.* 59 (17): 8295–8297.

Carriger, J.F., Barron, M.G., and Newman, M.C. (2016). Bayesian networks improve causal environmental assessments for evidence-based policy. *Environ. Sci. Technol.* 50 (24): 13195–13205.

CDFW [California Department of Fish and Wildlife]. 2020c. Spring Kodiak Trawl. Available from http://www.dfg.ca.gov/delta/projects.asp?ProjectID=SKT (accessed 20 May 2020).

CDPR [California Department of Pesticide Regulation]. 2019. Surface Water Database (SURF). https://www.cdpr.ca.gov/docs/emon/surfwtr/surfdata.htm (accessed 3 December 2019).

CDWR [California Department of Water Resources]. 2020b. CNRA Open Data Home. https://data.cnra.ca.gov/organization/dwr/portal/home (accessed 21 May 2020).

CDWR [California Department of Water Resources]. 2020a. DWR Atlas. http://atlas-dwr.opendata.arcgis.com/datasets/legal-delta-boundary.

Dillon, T., Beckvar, N., and Kerns, J. (2010). Residue-based mercury dose-response in fish: an analysis using lethality-equivalent test endpoints. *Environ. Toxicol. Chem.* 29: 2559–2565.

Duncan, P.B. (2005). Interactions among risk assessors, decision makers and stakeholders at the regional scale: the importance of connecting landscape level endpoints with management decisions. In: *Regional Scale Ecological Risk Assessment* (ed. W.G. Landis), 37–52. CRC Press.

Graham, S.E., Chariton, A.A., and Landis, W.G. (2019). Using Bayesian networks to predict risk to estuary water quality and patterns of benthic environmental DNA in Queensland. *Integrated Environmental Assessment and Management* 15 (1): 93–111.

Harris, M.J., Stinson, J.M., and Landis, W.G. (2017). A Bayesian approach to integrated ecological and human health risk assessment for the South River, Virginia mercury contaminated site. *Risk Anal.* 37: 1341–1357.

Hart Hayes, E. and Landis, W.G. (2004). Regional ecological risk assessment of a nearshore marine environment: Cherry Point, WA. *Hum. Ecol. Risk. Assess.* 10: 299–325.

Hart Hayes, E. and Landis, W.G. (2004). Regional ecological risk assessment of a nearshore marine environment: Cherry Point. *WA. Human and Ecological Risk Assessment.* 10: 299–325.

Herring, C.E., Stinson, J., and Landis, W.G. (2015). Evaluating Non-Indigenous Species Management in a Bayesian Networks Derived Relative Risk Framework for Padilla Bay. *Washington. Integr Environ Assess Manag.* 11: 640–652.

Johns, A.F., Graham, S.E., Harris, M.J. et al. (2017). Using the Bayesian network relative risk model risk assessment process to evaluate management alternatives for the South River and Upper Shenandoah River, Virginia. *Integr. Environ. Assess. Manag.* 13: 100–114.

Kaikkonen, L., Parviainen, T., Rahikainen, M. et al. (2021). Bayesian networks in environmental risk assessment: a review. *Integr. Environ. Assess. Manag.* 17: 62–78.

Laetz, C.A., Baldwin, D.H., Collier, T.K. et al. (2009). The synergistic toxicity of pesticide mixtures: implications for risk assessment and the conservation of endangered pacific salmon. *Environ. Health Perspect.* 117: 348–353.

Landis, W.G. (2002). Uncertainty in the extrapolation from individual effects to impacts upon landscapes. *Human and Ecological Risk Assessment.* 8: 193–204.

Landis, W.G. (2003). Twenty years before and hence; ecological risk assessment at multiple scales with multiple stressors and multiple endpoints. *Human and Ecological Risk Assessment.* 9: 1317–1326.

Landis, W.G. (2005). *Regional Scale Ecological Risk Assessment Using the Relative Risk Model*, 286. Boca Raton: CRC Press Edited Book.

Landis, W.G. (2021). The origin, development, application, lessons learned, and future regarding the Bayesian network relative risk model for ecological risk assessment. *Integr. Environ. Assess. Manag.* 17 (1): 79–94.

Landis, W. and Wiegers, J. (1997). Design considerations and a suggested approach for regional and comparative ecological risk assessment. *Hum. Ecol. Risk. Assess.* 3: 287–297.

Landis, W. and Wiegers, J. (2005). Introduction to the regional risk assessment using the relative risk model. In: *Regional Scale Ecological Risk Assessment Using the Relative Risk Model* (ed. W. Landis), 11–36. Boca Raton, FL: CRC Press.

Landis, W.G., Duncan, P.B., Hart Hayes, E. et al. (2004). A regional assessment of the potential stressors causing the decline of the Cherry Point Pacific herring run and alternative management endpoints for the Cherry Point Reserve (Washington, USA). *Human and Ecological Risk Assessment* 10: 271–297.

Landis, W.G., Hart Hayes, E., and Markiewicz, A.M. (2005). Chapter 12. Retrospective Regional Risk Assessment Predictions and the Application of a Monte Carlo Analysis for the Decline of the Cherry Point Herring Stock. In: *Regional Scale Ecological Risk Assessment Using the Relative Risk Model* (ed. W.G. Landis), 245–256. Boca Raton: CRC Press.

Landis, W.G., Ayre, K.K., Johns, A.F. et al. (2017a). The Multiple Stressor Ecological Risk Assessment for the Mercury Contaminated South River and Upper Shenandoah River Using the Bayesian Network-Relative Risk Model. *Integr Environ Assess Manag.* 13: 85–99.

Landis, W.G., Markiewicz, A.J., Ayre, K.K. et al. (2017b). A general risk-based adaptive management scheme incorporating the Bayesian network Relative Risk Model with the South River, Virginia, as case study. *Integr Environ Assess Manag.* 13 (115-126): 2023.

Landis, W.G., Chu, V.R., Graham, S.E. et al. (2020). Integration of chlorpyrifos acetylcholinesterase inhibition, water temperature, and dissolved oxygen concentration into a regional scale multiple stressor risk assessment estimating risk to Chinook salmon. *Integr. Environ. Assess. Manag.* 16: 28–42.

Landis, W.G., Lawrence, E.J., Bowers, M. et al. (2024a). *The Relative Contributions of Contaminants to Ecological Risk in the Upper San Francisco Estuary*, 14. Institute of Environmental Toxicology & Chemistry Publications.

Landis, W.G., Mitchell, C.J., Hader, J.D. et al. (2024b). Incorporation of climate change into a multiple stressor risk assessment for the Chinook salmon (*Oncorhynchus tshawytscha*) population in the Yakima River, Washington, USA. *Integr. Environ. Assess. Manag.* 20 (2): 419–432.

Marcot, B.G., Steventon, J.D., Sutherland, G.D., and McCann, R.K. (2006). Guidelines for developing and updating Bayesian belief networks applied to ecological modeling and conservation. *Canadian Journal of Forest Research* 36: 3063–3074. https://doi.org/10.1139/x06-135.

Mitchell, C.J., Lawrence, E.J., Chu, V.R. et al. (2021). Integrating metapopulation dynamics into a Bayesian network relative risk model: assessing the risk of pesticides to Chinook salmon (*Oncorhynchus tshawytscha*) in an ecological context. *Integr. Environ. Assess. Manag.* 17: 95–109.

Nyberg, J.B., Marcot, B.G., and Sulyma, R. (2006). Using Bayesian belief networks in adaptive management. *Can. J. For. Res.* 36: 3104–3116.

Pollino, C.A., Woodberry, O., Nicholson, A. et al. (2007). Parameterization and evaluation of a Bayesian network for use in an ecological risk assessment. *Environ Model Software* 22: 1140–1152.

Ritz, C., Baty, F., Streibig, J.C., and Gerhard, D. (2015). Dose-response analysis using R. *PLOS ONE* 10 (12).

SFEI and ASC (San Francisco Estuary Institute and The Aquatic Science Center) (2019). California environmental data exchange network (CEDEN). https://www.sfei.org/projects/california-environmental-data-exchange-network-ceden (accessed 3 December 2019).

USEPA (United States Environmental Protection Agency) (2020a). Aquatic life benchmarks and ecological risk assessments for registered pesticides. USEPA Office of Pesticide Programs. https://www.epa.gov/pesticide-science-and-assessing-pesticide-risks/aquatic-life-benchmarks-and-ecological-risk (accessed 28 April 2020).

USFW (U.S. Fish and Wildlife) (2020). *Delta Juvenile Fish Monitoring Program*. Lodi, CA: U.S. Department of Interior. https://www.fws.gov/project/delta-juvenile-fish-monitoring-program (accessed May 8 2026.

Uusitalo, L. (2007). Advantages and challenges of Bayesian networks in environmental modelling. *Ecol Model.* 203: 312–318.

Wiegers, J.K., Feder, H.M., Mortensen, L.S. et al. (1998). A regional multiple stressor rank-based ecological risk assessment for the fjord of Port Valdez, AK. *Hum. Ecol. Risk. Assess.* 4: 1125–1173.

8

Integrating Emerging Science in Advancing Risk Assessment for Wildlife

Mark S. Johnson[1] *and Thomas G. Bean*[2]

[1] *Defense Centers for Public Health – Aberdeen, Toxicology Directorate, Aberdeen Proving Ground, MD, USA*
[2] *FMC Corporation within Global Regulatory Sciences, Newark, DE, USA*

Introduction

Assessing risks to wildlife from exposures to chemicals has a fundamental and conceptual basis founded in past releases and adverse effects that subsequently resulted in regulation that now requires it. Examples where harm to wildlife populations is best illustrated by effects from environmental releases of DDT that biomagnified in the food chain and dramatically affected reproductive success of upper trophic-level birds (Peakall 1970; Rattner 2009). Populations of many raptor species in North America (and elsewhere) were affected that resulted in extirpation of many species in some regions.

Many other examples exist where anthropogenic contamination of the environment has resulted in adverse effects on wildlife. Persistent chemicals such as polychlorinated biphenyls (PCBs), metals, and organochlorine pesticides have been shown to result in adverse effects to wildlife species (Aulerich et al. 1974; Hickie et al. 2007; Rattner 2009).

Many jurisdictions require the probability for adverse effects to wildlife to be estimated for new pesticides and biocides, and chemicals from industrial release and from environmental contamination. This process of assessing risk to wildlife is founded in methods used to evaluate risks to humans where exposure is estimated and compared to a toxicity-based criterion (a benchmark). This method does not provide a probability estimate of risk but rather is used within a phased approach to screen for risk and is initially biased toward protection in both the estimate for exposure and in the derivation of the toxicity-based benchmark. This is commonly known as the hazard quotient method.

This hazard quotient method has been criticized by many that it often does not provide measures of risk (probabilities) and has few actual data that support its utility (Tannenbaum et al. 2003; Sorensen et al. 2004). However, the logic supporting the hazard quotient paradigm is sound. Issues arise when multiple assumptions are used to estimate exposure and toxicity and further efforts to refine estimates are focused solely on providing greater precision in the exposure and toxicity side of this simple equation. Ecological interactions often go beyond an individual-focused assessment, where significant and profound consequences from environmental contamination can occur to prey species or habitat that are often not a part of this equation (Gibbons et al. 2015; Saaristo et al. 2018).

In recent years, many new methods have emerged from the advancing of science in ecology, healthcare, and biology. Advances in measuring genetic responses to exposures in vitro and in vivo have expanded exponentially as well as our understanding of immune function, neurology, and endocrine disruption. Computer models that estimate toxicity from chemical structures have emerged and are now commonly used for aquatic species when data are lacking but are not currently available for wildlife. Wildlife genomic interactions and responses have shown to be useful in conservation (Romanov et al. 2009). Isolation of immune cells from exposed wild birds has shown utility in observing immunosuppression (Grasman 2002). Wildlife cell lines have been developed and used when toxicity targets are suspected (Matson et al. 2009; Lee et al. 2018). Although the availability of wildlife-specific cell lines are increasing, they are rarely used to address uncertainty in estimating risk.

Methods to better estimate the bioaccessibility of chemicals in the environment have greatly improved. However, many of these new tools lack specific applicability to terrestrial vertebrate wildlife species. Models have been developed (and tested) that better emulate the movements of wildlife species that can demonstrably improve estimates of exposure (Johnson et al. 2007, 2009) yet are rarely used. Our ecological knowledge of factors that influence the density of species in a metapopulational context is also increasing (Verboom et al. 2001; Bélisle and Desrochers 2002; Dolman et al. 2007; Carlsen et al. 2009). Here, we provide a brief review of some of these emerging methods and provide specific suggestions to improve decision-making regarding exposures of anthropogenic chemicals (xenobiotics) to terrestrial vertebrate wildlife.

Despite these advances in our knowledge, very few of them have been implemented in a risk assessment context. We discuss some more recent advances in policy and point out some existing regulatory frameworks where emerging science in improving risk estimates for wildlife have application. Also provided are suggestions for further research into these areas.

Considering Data Generated from New and Emerging Tools in Problem Formulation

Clearly describing the problem is the first step in wildlife risk assessment. The problem formulation phase requires a clear understanding of the question and precisely how data will inform a decision once addressed.

Data collected by using new methods requires an explicit description of how those data will be used in advance to include, given a range of outcomes, how decisions could be made. Some considerations are provided as follows.

a) Conceptual Model Applications
b) Considering Data Quality Objectives
c) Integrating Model Output in Decision-Making

Consistent with the outcome of a recent workshop on this topic, the problem formulation stage involves consideration of integrating data for the use of new models and methods and developing protection goals considering the range of potential outcomes to be used as decision criteria (Sample et al. 2022). Many challenges were provided, and possible solutions integrating new methods were proposed. The rate of emerging science relevant to improving risk predictions is increasing, requiring risk assessors and managers to be familiar with the conceptual basis for its use.

Improved Methods for Exposure Estimation

Exposure (i.e., the dose) is integrated with the hazard endpoint to estimate risk. Typically, oral exposures are considered most relevant to terrestrial wildlife species; however, often little is known regarding interspecies differences in exposure via dermal and inhalation routes, whether it be based on behavioral (e.g., dust bathing in birds) or biochemical (e.g., dermal absorption potential). It is recommended that chemical physicochemical properties affecting fate and transport be used when prioritizing and estimating pathways for exposures, and the conceptual models should be consulted.

There are a host of ecological, behavioral, and life-history aspects that have the potential of greatly affecting exposure (Morrissey et al. 2023). Birds, in particular, have unique physiologies (e.g., respiratory and reproductive kinetics), complex behavioral responses that include transient territoriality, high site fidelity, and seasonal niche partitioning and are highly vagile (i.e., move freely and easily from place to place) where some species regularly migrate thousands of kilometers biannually. Exposures include those encountered at wintering areas, breeding areas, and migratory stopover sites. At breeding and wintering areas, movement within these sites can be limited. Consideration of factors that may affect exposure are provided in Figure 8.1.

Exposures to ectoderms are also complex, where ambient temperatures can affect feeding behavior, movement, and metabolism. Many species of snakes feed irruptively where exposures result in pulsed experiences. There are relatively few studies that can provide species-specific recommendations for exposure characterization for many reptilian species.

To evaluate potential improvements to exposure models, it is first necessary to understand the types of approaches that are currently used (see Figures 8.2 and 8.3). To illustrate this, examples of exposure calculations used in pesticide risk assessments in the European Union and the United States are described.

The quality of exposure models is largely a factor of the quality of the available data. The inputs to these models are typically residues in food items, rates with which residues decline and information around food intake rate per unit body weight, dietary composition, and time spent in a contaminated area (e.g., European Food Safety Authority [EFSA] 2009, 2023; USEPA 2012). These approaches have remained largely unchanged for decades.

Risk assessments of pesticide applications in Europe for birds and mammals primarily assumes exposure to be from dietary sources and includes options for spray applications, granules, and seed treatments. Additionally, secondary poisoning for fish-eating and earthworm/invertebrate-eating birds and mammals is also considered (and now benthic invertebrate-eating birds and mammals will also be considered) if the chemical exceeds the threshold at which it is likely to bioconcentrate (Log_{kow} of $\geq$3). Exposure via contaminated drinking water is considered for birds, and in puddles it is considered for both birds and mammals.

The US Environmental Protection Agency (USEPA) have a similar approach to EFSA in terms of the general parameters used for exposure assessments, but the approach is less detailed and specific, thus having the advantages of being simpler and easier to use. It also does not include assessment of secondary poisoning.

The USEPA have used the latest version of T-REX (Terrestrial Residues Exposure, v1.5.2) since June 2013 for calculation of exposure (and risk quotients for birds and mammals) in a spreadsheet-based model (https://www.epa.gov/pesticide-science-and-assessing-pesticide-risks/t-rex-version-15-users-guide-calculating-pesticide, USEPA 2012). Initially, residue concentrations in the food item of interest are calculated assuming degradation by first-order kinetics using the following equation. The maximum residues are calculated for selected food items in daily intervals for a period of one year, and the maximum residue during that period is taken as the estimated environmental concentration (EEC).

$$C_t = C_0 e^{-kt}$$

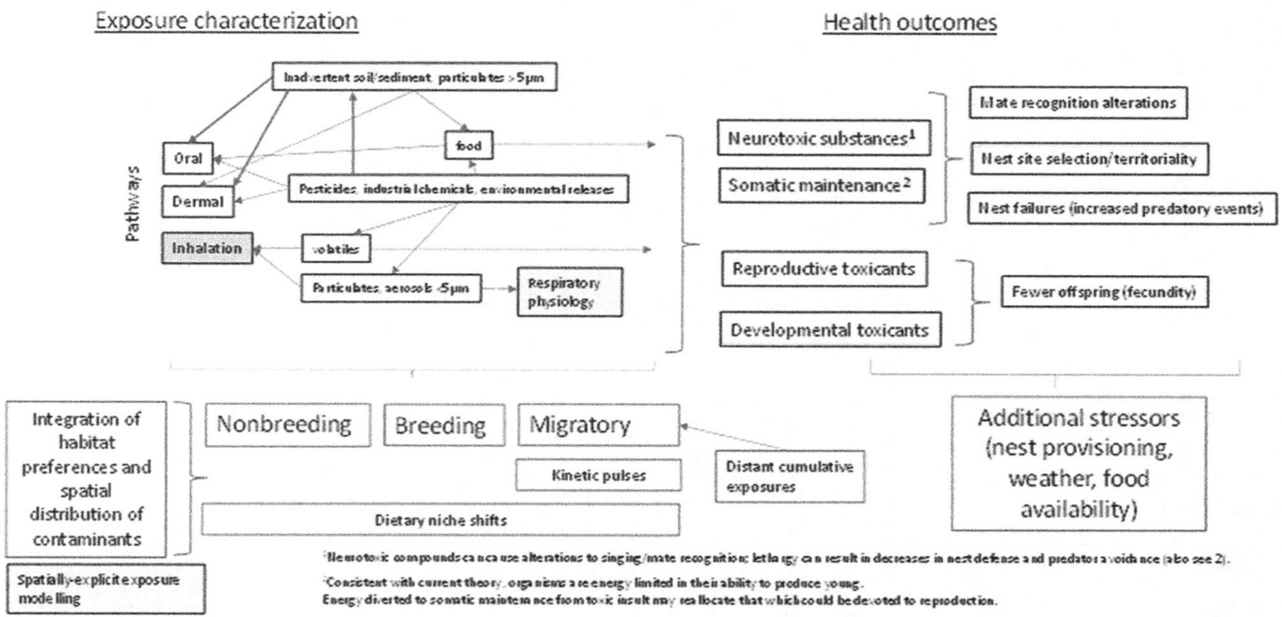

Figure 8.1 Conceptual diagram of factors potentially influencing exposure and effects in birds.

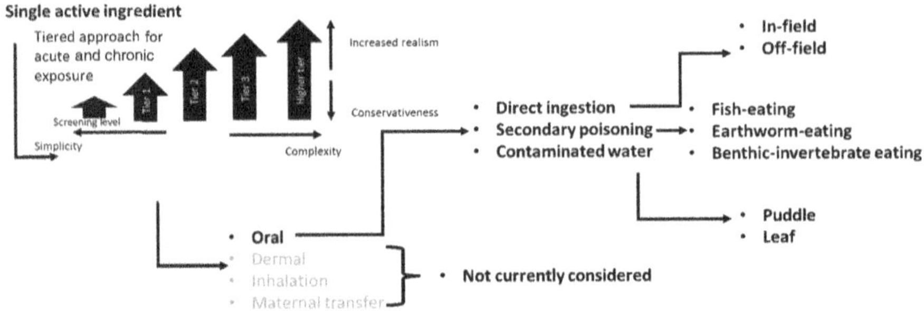

Figure 8.2 Simplified schematic of current approaches to pesticide exposure assessment for birds and mammals.

which can be expressed in natural log form as

$$\ln(C_t) = \ln(C_0) - kt$$

(Equation taken from USEPA 2012).

In this equation, C_t is the residue concentration in the crop at a given time. C_0 is the initial concentration, which is given in lb/acre and are multiplied by conversion factors for the 90th percentile (240 for short grass, 110 for tall grass, 135 for broad-leafed plants, 15 for fruits/pods, and 94 for arthropods) and median (85 for short grass, 36 for tall grass, 45 for broad-leafed plants, 7 for fruits/pods, and 65 for arthropods) residues, respectively. These conversion factors are based on the Kenaga nomogram from Hoerger and Kenaga (1972), which was modified by Fletcher et al. (1994). As previously mentioned, the 90th-Percentile Kenaga residues are to be used in risk assessment. T-REX can also calculate exposure following multiple applications. If multiple applications are used, then the residues from the final application and any remaining residues from the previous applications are summed and then entered into the first-order kinetic dissipation equations.

The residue concentrations are then combined with food intake rates for birds (20, 100, and 1,000 g, respectively) and mammals (15, 350, and 1,000 g, respectively) of different sizes, broadly categorized as small, medium, and large. Allometric scaling is used to calculate food intake rate for these using the following empirically derived equations based on data collected for the USEPA's Wildlife Exposure Handbook (USEPA 1993) to calculate consumption for birds and mammals. In these equations, F is the food intake rate (g fresh weight/kg BW), BW is body weight (g), and W is the proportion of the food items that is water (0.1 for granivores and 0.8 for herbivores and insectivores).

$$F = (0.648 \times BW^{0.651})/(1 - W) \text{ (Birds)}$$

$$F = (0.621 \times BW^{0.564})/(1 - W) \text{ (Mammals)}$$

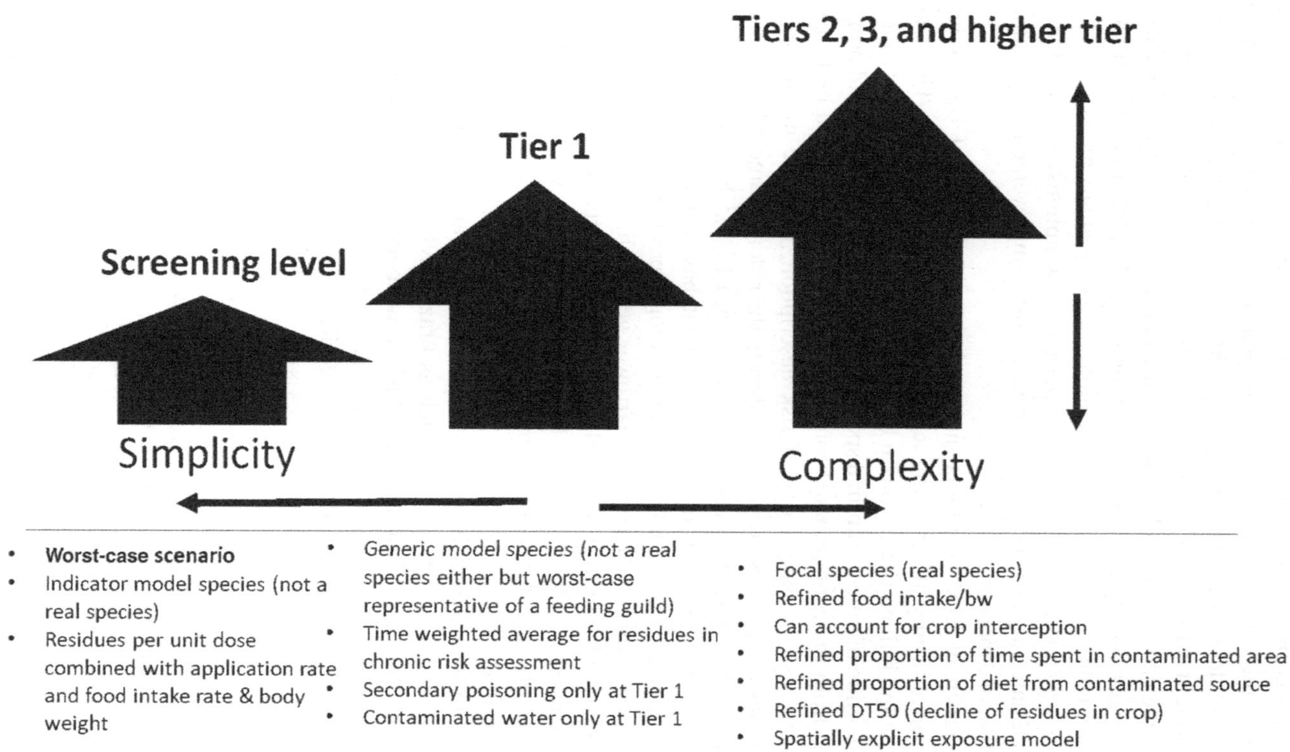

Figure 8.3 Simplified tiered approach for acute and chronic exposure to pesticides for birds and mammals.

USEPA's exposure calculation has not been updated in a decade, and the general framework has been the same for longer than that (no record of a date for version 1 of T-REX could be found, but version history in T-REX indicates that, many of the underlying data requirements for the residues and the food intake rate per unit body weight calculations are based on work that was published 30 years ago or more; some exposure calculations are based on default criteria from sources that are decades old [e.g., EFSA 2023's secondary poisoning and contaminated water calculations contain numerous examples of data that are 30–50 years old]). Revisiting and evaluating some of these underlying assumptions in light of today's changing climate and agricultural practices may offer improvements in accuracy of calculation of exposure concentrations.

The approach of EFSA contains more scenarios than that of USEPA with different tiers (starting at screening level then progressing through Tier 1 and higher tiers) and refinement options at higher tiers such as time-weighted average factor, proportion of time the animal spend in the contaminated area, and proportion of diet from the contaminated source that are not incorporated by USEPA. USEPA's approach is instead one of risk description, with the mean Kenaga nomogram (residue) values providing context to the risk assessment based on the 90th-percentile Kenaga values (Hoerger and Kenaga 1972).[1] EFSAs residues are based on a large residue per unit dose (RUD) database derived from field data collated from many RUD studies in different crop types (EFSA 2023) and thus may have more realism but perhaps narrower geographic applicability/limited applicability outside of European conditions. Improvements to exposure models need to balance the complexity of the European assessments with the simplicity and user-friendly spreadsheet-based model of USEPA's T-REX calculation tool (discussed as follows).

Currently, exposure models for species in other geographic regions are lacking (e.g., major agricultural markets such as Brazil). While regulatory agencies in Brazil (e.g., Brazilian Institute of the Environment and of Renewable Natural Resources [IBAMA]) are currently seeking to develop or adapt existing risk assessment frameworks for wildlife, the task of identifying relevant model species and gaining understanding of their life histories is one of many challenges that must be overcome. The differences in climatic conditions and year-round growing seasons mean assumptions around exposure are likely quite different to Europe, and so it

1 The Kenaga nomogram has been used by USEPA to estimate maximum residue levels (parts per million based on fresh weight) on day 0 following pesticide applications assuming a linear residue dose relationship at different rates since the early 1980s to one of six different plants or plant parts: short grass, long grass, leaves and leafy crops, forage, pod containing seeds, and fruits (Fletcher et al. 1994). The data used to develop the Kenaga nomogram came from a 1972 paper by Hoerger and Kenaga, hence the name "Kenaga nomogram". The dataset was based on 27 pesticides and 36 different crop types; however, in 2023 many of the pesticides upon which the data were developed are now banned.

would be better to invest resources to create Brazil-specific exposure models than to simply try to apply what has been developed by EFSA for screening and Tier 1 (as appears the current approach of IBAMA). Filling these data gaps is an extremely challenging (even laborious) task and will likely take many years to complete. In the following section, we review some of the new technologies and approaches that could be/are being incorporated into exposure assessments.

Aspirational Ways to Improve Accuracy of Exposure Models

As part of a SETAC workshop focused on providing suggestions to move wildlife risk assessment in line with the state of the science in the 21st century, an exposure workgroup published suggestions around four key areas of complexity that they concluded are not currently captured by present-day frameworks (Morrissey et al. 2023): chemical complexity, environmental complexity, organismal complexity, and ecological complexity.

Morrissey et al. 2023 stated, "If chemicals are inappropriately screened out at a lower tier, there is no opportunity to revisit the assessment using more comprehensive and realistic methods."

Chemicals that fail at the screening level can have risk assessments refined at higher tiers. The simplicity of the screening level should largely remain so that resources (of regulators and registrants) can be channeled to where they are most needed; otherwise, there is no purpose to an overzealous or ineffective screening tool. Indeed, the case studies provided in Morrissey et al. (2023) about integration of complexity in exposure assessments are examples of higher-tier refinements and not screening-level exposure assessments. It may be possible to integrate some elements of the matrix provided by Morrissey et al. (2023) at a screening level, e.g., combined oral, dermal, inhalation exposure, when exposure models are available and validated for dermal and inhalation routes. Efforts to incorporate new science (i.e., the four areas of complexity not currently incorporated that Morrissey et al. 2023 highlighted) should be integrated as higher-tier refinement options to more accurately characterize exposure rather than add unnecessary complexity to a worst-case exposure scenario at the screening level. Morrissey et al. (2023) provided a checklist of items that could be incorporated into exposure assessment for each of the four areas. In Figure 8.4, annotations have been added to highlight where these areas could slot into an existing approach based on the EFSA-tiered approach to Exposure assessment presented in EFSA (2023).

Incorporating chemical complexity into exposure assessments could include consideration of mixtures, metabolites, and transformation products. At present, active ingredients and ecotoxicologically relevant metabolites are considered in

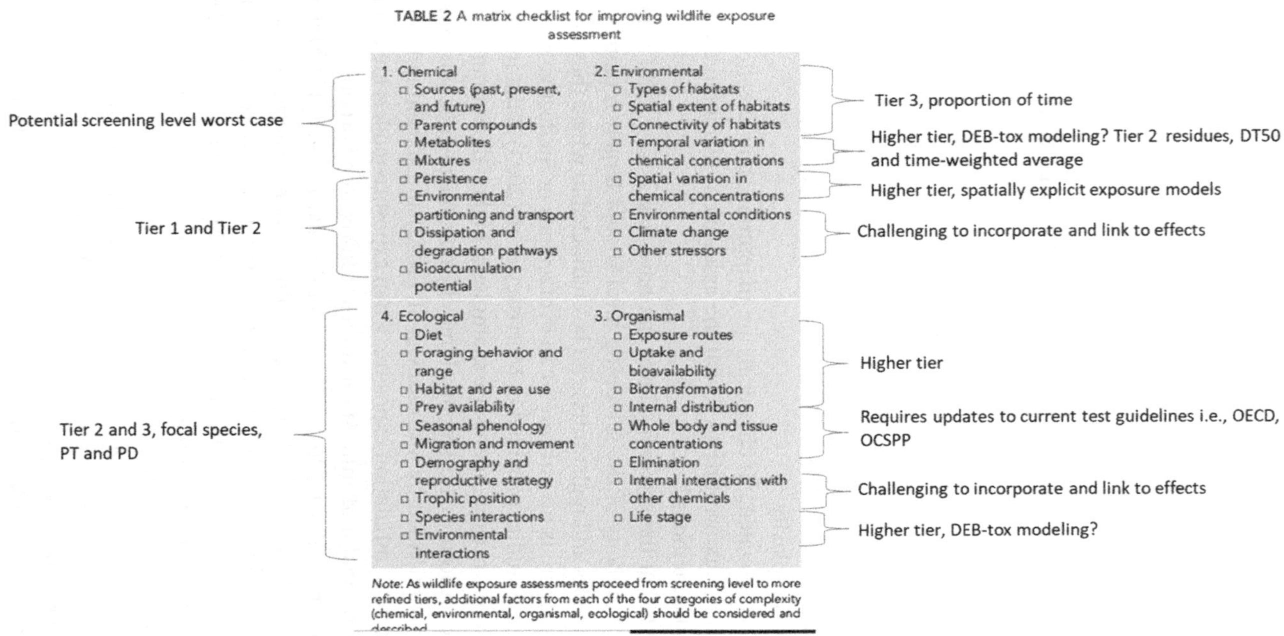

Figure 8.4 Schematic building upon ideas presented in Morrissey et al. (2023) to incorporate chemical, environmental, organismal, and ecological complexity into exposure assessment based on European Food Safety Authority (2023) exposure assessment tiers. Note Table 2 is taken from Morrissey et al. (2023).

isolation in pesticide risk assessments in the European Union, with an arbitrary additional assessment factor of 10 applied in the risk assessment. Adverse outcome pathways can be used to classify chemicals by mode of action, which could help integrate exposure for mixtures from within a chemical class.

Approaches such as the eco-exposome could be used to evaluate the totality of exposure of an organism to chemicals over the course of its lifetime (Morrissey et al. 2023). Dynamic energy budget (DEBtox) approaches (Jager et al. 2010) could play a role in accounting for time-variable exposures to different chemicals, e.g., EurMix (Beronius et al. 2020).

Environmental complexity includes incorporation of spatial and habitat heterogeneity with information on life histories (e.g., behavioral data) (see later in the chapter for further information on spatially explicit exposure models). Land use and landcover data are now widely available via satellite imaging and can be incorporated in exposure assessments for overlay with known distributions of endangered species. Conducting long-term biomonitoring instead of single time-point data will help improve understanding of fate and persistence of chemicals, which is important for long-term exposure assessment refinements for residue concentrations. Inclusion of environmental complexity in exposure assessment will likely become increasingly important as climate change alter species distributions as well as the timing, availability, and distribution of food items (Morrissey et al. 2023).

Organismal complexity includes approaches for linking external exposure to internal exposure. Toxicokinetic and toxicodynamic (TK/TD) approaches can help to predict absorption, distribution, metabolism, and excretion (ADME) parameters or effects associated with a given dose. Physiologically based kinetic (PBK) modeling can use information on physiology to back-calculate an external dose (reverse dosimetry) based on internal (tissue residues) from biomonitoring data (e.g., Scanes et al. 2021; Baier et al. 2022). At present, it is challenging to validate TK/TD and PBK approaches because the regulatory ecotoxicology studies do not require analysis of residues in tissues, so it means that internal residues are only available from the metabolism studies conducted in livestock and hens. in vitro approaches offer options to evaluate bioaccumulation and biotransformation, e.g., microsomes and hepatocytes (e.g., Kuo et al. 2022); however, connecting these approaches to exposure assessment into existing regulatory frameworks will likely require a paradigm shift in current risk assessment process. Techniques that account for bioaccessibility of food items (e.g., physiologically based extraction (Pbets) could help improve accuracy of internal exposure estimates). However, Pbets for wildlife have mostly been developed for metals and inorganics (e.g., Beyer et al. 2016), and few have been attempted for organics (e.g., Bean et al. 2016). At present, the focus of risk assessments is generally on oral exposure via diet, and to some extent via contaminated water. Holistic assessments that integrate not just dietary routes but also maternal transfer and dermal and inhalation exposure would provide a more comprehensive assessment, but care would need to be taken to avoid overly conservative scenarios that could not occur simultaneously at higher tiers, i.e., such conservativeness should be reserved for the screening level.

Incorporation of **ecological complexity** into higher-tier exposure assessment could include incorporation of life-history traits such as longevity, age at maturity, information on reproductive effort; ecological traits such as foraging behavior; and physiological traits such as thermoregulation, basal metabolic rate, body condition, and lipid content. Some information is available online on such traits (e.g., http://animaltraits.org; Herberstein et al. 2022); however, for the data to be useful for an exposure assessment, it will be needed for a diverse range of species and geographies. Information on metabolism, morphometrics, lipid content, body temperature, AVONET, and animal traits database (Tobias et al. 2022). Movements of animals, particularly migratory species, can be tracked using many methods, e.g., radio tracking devices, Motus nanotags, light level loggers, GPS, cellular, solar-powered satellite tags, and remote-sensing radar data (Morrissey et al. 2023). These data could be used to better inform the proportion of time estimates that individuals spend in the contaminated area, a refinement currently available at higher tiers in the European Union's risk assessment for pesticides. Time-variable exposure over the course of an animals' life could be captured by DEBtox modeling approaches to better incorporate factors such as seasonal shifts in diet and shifts in exposure due to life stage and reproductive stage.

Incorporation of these four areas of complexity into exposure assessment will demand resources that are already stretched thin for both registrants, consultants and regulators. The focus for improvements to exposure assessment should be to provide clear guidance on how to incorporate tools that add chemical, environmental, or ecological complexity in proportion to the problem faced without adding unnecessary unworkable requirements (i.e., it is not realistic for all elements of the matrix in Figure 8.4 to be integrated at a single assessment tier).

What Has Been Incorporated Recently into Exposure Models?

Some of the key changes to the exposure models used by ESFA in their 2023 guidance include making the secondary poisoning calculations more conservative than the previous version from 2009 with a reduction in the body weight of the generic model species and a reduction in the averaging period for the timeweighted average from 21 to 7 days. The underlying data used for the parameters in the exposure calculation, such as a new and much larger database for Residues per Unit

Dose (RUD) data, which helps improve the accuracy of the exposure calculation (e.g., incorporating data from Ebeling and Wang 2018). Another major change was adjusting the criteria for when the time-weighted average factor can be used to refine exposure in the long-term risk assessment (that are based on whether the effect endpoint is considered to be the result of short-term or long-term exposure); thus, the nature of the exposure is being connected to the effect/s of concern for the long-term risk assessment.

The new EFSA guidance document also includes exposure assessments for many new focal species (exposure calculations based on information on feeding guild/dietary composition, food intake rate per unit body weight), but also includes new exposure scenarios, such as benthic invertebrate-eating birds and mammals as well as consideration of risk for mammalian species that are not in the crop of concern but could be exposed via drift in an off-field exposure.

Exposure refinement options exist in terms of proportion time (PT, the proportion of the species time spent in the area where they could be exposed; e.g., Ludwigs et al. 2017) and proportion diet (PD) composed of the contaminated food items (e.g., Ludwigs et al. 2022). These require robust field data and may involve techniques such as radiotracking, GPS, and capture-mark-release-recapture methods for a relevant focal species, but in order for the studies to be considered reliable the studies must meet the requirements of the Critical Appraisal Tool (EFSA e.g., Appendix F of EFSA 2023; Lahr et al. 2023 for RUD studies). In the recent 2023 update, the criteria for the selection of the focal species was changed from the 2009 version of the guidance. Previously, the focus was on the most abundant species in the contaminated area; now, the focus is on species that may be "vulnerable" based on food intake rate per unit body weight, and they use the crop in a way that increases likelihood of exposure (i.e., PT and PD).

Future Options for Improving Exposure Assessment Models

EFSA's 2023 guidance document included some options for Tier 3 and higher-tier approaches that are not currently available but could be validated/available before a future update to the guidance document is issued. This includes identification of region-specific focal species for refinement of fish consumption in secondary poisoning assessments and Food web modeling (EFSA 2023). Food web modeling approaches could add dimensions to secondary poisoning exposure assessments that are not currently possible such as consideration of effects on community structure and links to ecosystem structure and function (EFSA 2023). At present, Food web models are not as well developed for terrestrial systems as they are for aquatic systems. At present, food web models are primarily focused on diffusion transport systems in lipid (Gobas et al. 2016) and are lacking for chemicals that are not lipid stored. Additionally, datasets for developing and validating wildlife food web models (e.g., transfer from air and soil to herbivorous and carnivorous organisms) are lacking. Generation and development of such data could help to further the use of food web models in wildlife risk assessments.

While these approaches to exposure assessment model are focused on field observations in free-ranging individuals and populations, another option for improvement of exposure models is to incorporate models that enable better links to be made between external exposure with internal exposure (and subsequent effects). For example, this might include the incorporation of PBK models that consider ADME rates and provide estimates of internal and target organ dose that can reduce the need for in vivo testing of other species where there is evidence of biological conservance of toxicological mechanisms (e.g., Scanes et al. 2022). While there has been progress in PBK for mammals in recent years (due to application in human health risk assessment), work for birds and wildlife species more generally has not made the same steps forward (Scanes et al. 2022). While there are similarities between birds and mammals for some parameters such as hemoglobin concentrations, there are marked differences for others such as relative weight of kidney, spleen, and ovary (Scanes 2022). A significant challenge with developing PBK models for birds is the differences in physiology between males and females especially when in reproductive physiologies. Baier et al. (2022) developed and parameterized an avian PBK model for northern bobwhite (*Colinus virginianus*), mallard (*Anas platyrhynchos*), and chicken (*Gallus gallus*) and compared model output for nine chemicals. Compared with the in vivo data, 22–79% of the model simulations were within a factor of 3 and 41–89% were within a factor of 10.

From the perspective of utility of predictions of internal residues, it is unfortunate that the current avian reproduction studies, OECD 206 (OECD 1984), and OCSPP 850.2300 (USEPA 2012b) do not include requirements to collect tissues for residue work (just a preferred request to do residue work in OCSPP 850.2300 for chemicals unless elimination is demonstrated to be <24 hours). EFSA's 2023 guidance document discusses the potential for TK/TD models (of which PBK models are essentially physiologically based toxicokinetic models) to predict mortality and sublethal effects by better relating external and internal exposures. However, while these models hold great potential for achieving this goal at some point in the future, they are not available yet, and validation without residue data from the existing studies conducted to evaluate effects is a shortcoming. Indeed, EFSA include the following recommendation: "Therefore, the working group advises investigation of the possibility of adapting these tests so that test results (e.g., additional ADME information and pesticide dynamics in blood samples) can be optimally used for TK and/or TKTD modelling purposes in the near future, while keeping the number of test animals at a minimum."

The differences in objectives of a TK/TD and an avian toxicity test mean that it will be challenging to add sampling at intervals other than termination without the addition of a satellite group. Recent work by Trijau et al. (2023) has demonstrated the potential application of Dynamic Energy Budget (DEB)–TK/TD models to link effects observed on egg laying in standard toxicity tests with TK/TD modeled internal residues. The model was validated for five pesticides for two species (northern bobwhite and mallard) against in vivo data and was able to differentiate between effects caused as a result of toxic action of the chemical and food avoidance (Trijau et al. 2023). Such modeling approaches hold potential application to reduce in vivo testing in the future, but there is still a long way to go. In an avian reproduction study, there are typically 15 endpoints evaluated and the DEB-TKTD model has so far only been applied to the endpoint of egg production where the extrapolation from the exposure (to the adult birds) is shortest.

- **Spatially explicit exposure models.** Spatially explicit models have shown utility for predicting risk, particularly with terrestrial wildlife species that are vagile (Hope 2002; Johnson et al. 2007; Purucker et al. 2007; von Stackelberg et al. 2017); some can include aspects of multiple stressors in the analysis (Hope 2004). Spatial resolution of chemical concentration and habitat preferences must be used to achieve protective yet accurate results (Hope 2000; Johnson et al. 2007). Spatially explicit, individually based models often provide output in the form of mean daily exposure estimates (mg/kg-d) for a user-defined number of individuals in a population. These mean values can be compared to a TRV to develop a curve of population-level risk (Figure 8.5).

Spatial extent of chemical contamination is user defined and can be user drawn on a map or extracted from geographical information system (GIS) applications. It can be developed as general as a high, medium, and low where statistical methods can be used to extrapolate data from actual environmental samples (e.g., 95% upper confidence level or highest sample concentration for each of the three levels). Habitat suitability polygons represent the second layer of spatial data and can be constructed from habitat suitability indices (Dijak et al. 2007) or estimated based on species-specific habitat requirements. Food web models occur at each foraging event, and bioaccumulation factors are also user defined and can be based on site-specific criteria to include inadvertent soil ingestion. Daily foraging ranges should be determined from literature value or based on site-specific observations based on seasonal life-history information. Changes in habitat suitability is a unitless number that ranges from 0 (no habitat value) to 1 (high value) that influences exposure time in specific areas.

Spatially explicit exposure models are also useful for conducting feasibility estimates for decision-making. Risk assessors can ask "what if" questions regarding

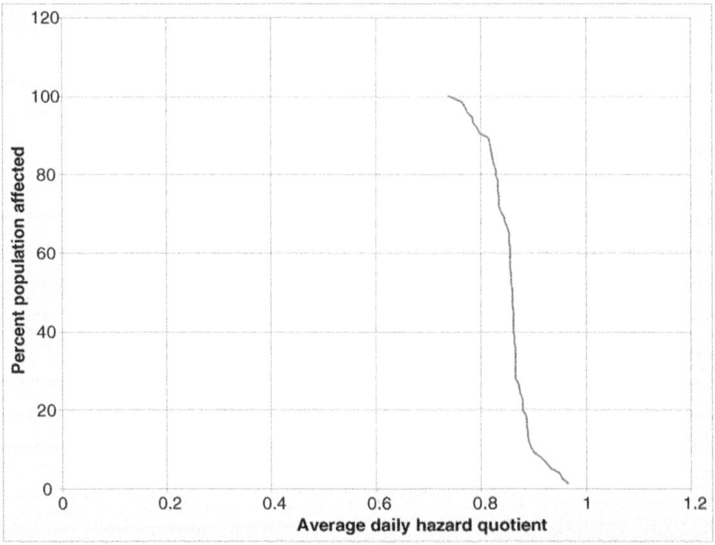

Figure 8.5 Individual mean ratio of exposure of lead in soil to Eastern Bluebirds compared to the EcoSoil screening level lead toxicity reference value (1.63 mg/kg-d) for birds for a lead exposure within a small arms range. Model assumed 15 foraging event/day and a population size of 16 birds for a nesting season of 150 days.

the influence of contamination by virtually remediating hot spots to determine if space matters in reducing risk. The utility of this function cannot be overestimated as space clearly matters in regard to exposure when habitat preferences are considered.

Although spatially explicit exposure models provide greater precision than deterministic models, there can be considerable effort involved in setting them up and typically can only function for one substance at a time. Furthermore, there is little evidence that there is a benefit when using spatially explicit exposure models for species with small daily foraging ranges such as small mammals (Johnson et al. 2020). They can, however, be quite sophisticated and include integration of other stressors (Duggan et al. 2015). Scenario modeling has been suggested as way to augment traditional chemical risk assessment to enable the integration of various nonchemical stressors (e.g., land management options, habitat alterations, stochastic human interference; Harwood 2000; Hobday et al. 2011). Many of these multistressor risk assessments attempt to characterize and account for the uncertainties from measurement and estimation error, imperfect model parameters, and simple lack of knowledge. Although examples of incorporating effects of multiple stressors exist for fisheries and ecosystem effects in aquatic environments, we have not found any examples that include terrestrial wildlife.

Novel Tools for Estimating Hazard (Toxicity)

Increasingly, ecological data provide evidence that suggests effects outside mortality, growth, and reproduction is important in population maintenance (Sih et al. 2012; Bertram et al. 2022). Toxic exposures to some substances can cause behavioral alterations (e.g., lethargy; see, for instance, Hussain et al. 2008), particularly to pesticides and neurotoxicants (Mitra et al. 2021). Furthermore, changes in behavior can be an important factor in extirpations of bird populations (Reed 2001). Data from the effects of chemical exposures using assays such as immortal neural cell lines that can predict neurological effects may have value in determining the probability for behavioral changes that could affect exposed individuals. Moreover, evaluating genomic responses within these cell lines may provide early indications of adverse effects and can have application in addressing questions regarding differences in toxicodynamics between species where biological pathways are conserved. Dose–response information may be useful in establishing toxicity reference values when considering toxicokinetics (TK) (e.g., in vitro/in vivo extrapolation).

Increasingly, the integration of mechanistic information (i.e., how exposure to a chemical causes adverse health effects on a molecular and biological level) is being used to assist in extrapolating between species (Davidson et al. 1986; Celander et al. 2011; van den Berg et al. 2021). The logic is that when biological pathways for adverse effects are conserved, extrapolating effects between species is supported. However, physiological differences (particularly in gastrointestinal structure) have the potential to greatly affect external and internal dose metrics (Celander et al. 2011; Johnson et al. 2010). Physiologically based pharmacokinetic models in wildlife hold promise to better consider species-specific differences in kinetics, particularly from oral exposures (Baier et al. 2022). Because species-specific partition coefficients are lacking for many species of wildlife, few PBPK models exist. These topics are further discussed in this chapter. Laboratory animal models are largely lacking for many classes of wildlife (or are sufficiently impractical for use). However, differences in physiologies can result in marked differences in TK (i.e., ADME) and therefore can greatly affect systemic dose. Many novel tools now exist to explore species-specific differences in toxicodynamics (e.g., new wildlife cell lines, mechanistic models, use of read-across), yet integrating the toxicodynamics and toxicodynamics for terrestrial wildlife species remains elusive.

In vivo *Alternative* Tools (Adverse Outcome Pathways, Cell Lines, IVIVE)

In vitro tools for wildlife risk assessment are currently not as well advanced as they are for human health risk assessment (e.g., Rattner et al. 2023). in vitro tools look to evaluate effects at lower levels of biological organization (i.e., molecular, cellular) and thus their utility to be applied to wildlife risk assessment is contingent upon them being biomarkers of downstream effects with significant consequences for individuals and populations. The use of quantitative PCR technology such as EcoToxChips (e.g., Basu et al. 2019) could have potential application in the screening stage of new chemical discovery. EcoToxChips have been developed and validated for two avian species (Japanese quail and double-crested cormorant) to evaluate effects of chemical exposure on transcription of ~380 toxicologically relevant genes (Basu et al. 2019; Farhat et al. 2019; Crump et al. 2020). In silico tools such as QSAR are not yet available for prediction of hazard thresholds in wildlife, but correlative approaches that use read-across techniques to extrapolate between species have been developed, e.g., WEB-ICE (Interspecies Correlation Estimation; Raimondo et al. 2015) with potential application for development of species sensitivity distributions. However, it must be stated that it has been the experience of the authors that it becomes difficult to determine if variability in response (toxicity) is related to species differences or those from study design (e.g., vehicle, bolus dosing, feeding studies). Previously discussed DEB-TKTD, PBK, and population models are available and validated to varying degrees for birds, but useful integration into risk assessment remains elusive (see Bean et al. 2023; Rattner et al. 2023).

In vitro assays for assessment of bioaccumulation and biotransformation of chemicals (e.g., microsomal enzyme activity, investigation of metabolic pathway, reviewed in Kuo et al. 2022) could have some application for identifying chemicals that are likely to bioaccumulate in terrestrial vertebrates and may thus be a concern for long-term risk assessment. However, on its own, avian bioaccumulation assessment is not part of any regulatory framework that these authors are aware of, and, thus, the utility would be limited to the discovery of new chemicals or would require updated regulatory frameworks to serve as either a screening tool for chronic in vivo testing or to inform a modified design (e.g., reduction of the exposure phase before photostimulation in an avian reproduction study). Read-across and utilization of existing in vivo data collected for assessment of metabolism and residues, e.g., laying hen studies have also been proposed to have some potential utility in wildlife risk assessment (Ebeling et al. 2023) and could be used to inform residues to inject in an egg injection study (e.g., Heinz et al. 2009) if the 15 endpoints evaluated in an avian reproduction study are broken down into separate areas to be evaluated by different new approach methods/nonanimal methods ("NAMs"). Additionally, in 2022, an avian in ovo screening assay (AIOSA) focused on endocrine assessment was proposed as a new Organisation for Co-operation and Development (OECD) test guideline by Japan. The proposed guideline would use domestic chicken embryos (*Gallus gallus*) (with potential expansion to Japanese quail, *Coturnix japonica*) as a model for an endocrine disruptor screening assay that would sit at level three in the OECD Endocrine Disruptor Assessment conceptual framework to evaluate

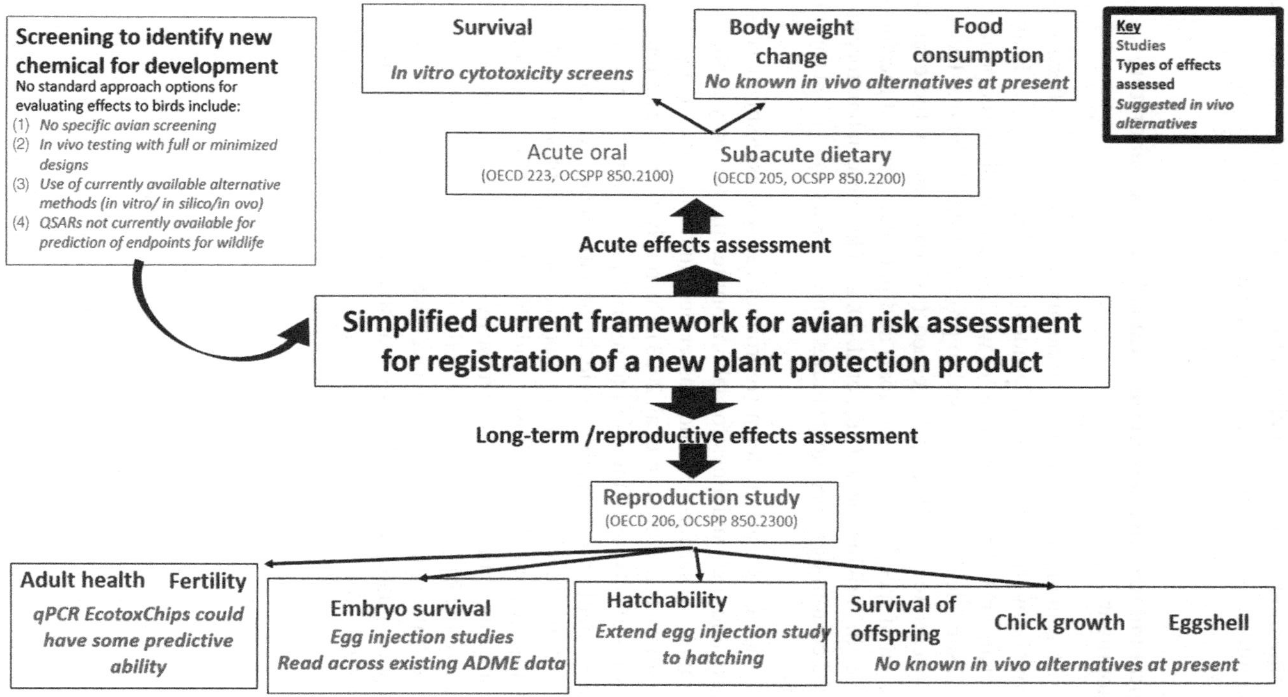

Figure 8.6 Schematic illustrating how in vivo alternative tools (in vitro, in silico, and in ovo) could complement existing hazard assessments for birds.

estrogenic, androgenic, antiestrogenic, and antiandrogenic effects. The schematic in Figure 8.6 illustrates how in vivo alternative tools (in vitro, in silico, and in ovo) could complement existing hazard assessments for birds.

Methodological Improvements in Current Guidelines and Laboratory Model Development

For better or worse, many wildlife risk assessments have focused on effects related to survival, growth, and reproduction for decades (see critical reviews in Bean et al. 2023; Rattner et al. 2023). Protection goals for wildlife are usually at the population level (although some legislation is focused to protect at the individual level, e.g., Endangered Species Act, 16 U.S.C. §1531 et seq. (1973)); however, there has been recent discussion of specific protection goals and potentially role of the "normal operating range" in deriving them (Galic et al. 2022). Therefore, laboratory guideline studies and custom-designed field studies tend to focus on marked (or frank) adverse effects.

Typically, an acute and a long-term (reproductive) risk assessment is made, but frameworks and test guidelines for laboratory testing are only available for birds (Galliformes, Anseriformes, and passerines) and mammals (rodents and lagomorphs) (summarized in Bean et al. 2023). Currently, there are no regulatory test guidelines for effects assessments in terrestrial phase amphibians (although endocrine assessments are typically made in aquatic phase amphibians) and reptiles (Bean et al. 2023), though some models and examples do exist (Bazar et al. 2010; Johnson et al. 2010; Johnson and Pillard 2023). The guideline tests available for mammalian testing are primarily conducted for human health assessment where the protection goals are at the individual level; thus, endpoints at molecular, cellular, tissue, and organ level are also assessed. For birds, the acute and subacute guidelines focus on survival and growth (body weight change) endpoints. Updates to regulatory test guidelines upon which laboratory testing protocols are based occur infrequently (see Bean et al. 2023 for a summary of the process involved in updating an OECD test guideline). Between October 2020 and October 2022, German Environment Agency (Umwelt Bundesamt, UBA) commissioned Ramboll and Fraunhoffer to conduct a review of existing OECD test guidelines (UBA 2023) with a view regarding the need for updates. It is notable that the avian reproduction study guideline received the most suggested updates of any guideline with 36 suggestions (OECD 206, OECD 1984). These included suggestions for the inclusion of a recovery phase, removal of measurement of food consumption in offspring, updates to numbers of replicates, and groups to better allow for effect concentration (e.g., ECx) estimation. Any updates to the mammalian guidelines would need collaboration with toxicologists/human health risk assessors and thus are more challenging and less likely to be updated than avian test guidelines.

Optimally, the determination of a threshold of toxicity is most useful for risk assessment. Typically, regulatory study design included three treatments and a control. If variation was relatively high in the toxicity endpoints measured, statistical differences between treatments are difficult to determine. Additionally, the thresholds defined are dependent on the doses selected for treatments in study design. Recently, EFSA 2023 recommended that the threshold for development of points of departure for long-term risk assessment should be the Benchmark dose (BMD) for a 10% difference from the control (BMD10) and not the No Observed Adverse Effect Concentration or Level (NOEC or NOAEL). The BMD method fits the best fit curve to the toxicity data (to include measures of variation), and it is far superior in determining the threshold for response. Unfortunately, the guideline methods for chronic studies have not been designed with BMD derivation in mind; it is unlikely that the dose–response relationship will be strong enough for regression-based statistical approaches for a test with only three dose groups and a control. Others have recognized this for contaminated sites and have bolstered toxicity study design to include at least four treatments and a control (Johnson and McAtee 2015). Animal welfare, global registration requirements, and space requirements in the laboratories that can conduct these tests mean that alternative designs to better support a BMD (by increasing the number of dose groups) may be challenging. Green et al. (2022) ran computer simulations to evaluate the ability of alternative designs that distribute the same number of adult animals (144) among more groups (6 groups of 12 pairs instead of 4 groups of 18 pairs). As the NOAEL can be used as the, (1) in countries outside of jurisdiction of the European Commission and (2) if a reliable BMD (as defined by EFSA's BMD guidance document (EFSA 2022)) cannot be derived. The alternative design improved the ability to generate a BMD (as more dose groups were included) and only reduced the power to detect a NOAEL or Lowest Observed Adverse Effect Level (LOAEL) by 12.5%. A study design with 12 pairs per group is acceptable under OECD 206 (OECD 1984) but not under USEPA's guideline OCSPP 850.2300 (USEPA 2012). Alternative designs with 16 pairs/group (the minimum required by USEPA are currently under investigation (e.g., Bean et al. 2022; Green et al. 2023)) and reducing the number of weeks of egg production from 10 to 8 weeks (and when the control group have met the validity criteria of laying ≥29 eggs/hen), which is acceptable under the guidelines, are also currently being evaluated. By reducing the number of weeks of egg production, it is possible to make significant savings on the number of animals used even if additional adult pairs are used as many of the animals used in the studies are the offspring. The number of adult pairs per group could not be reduced below 16 without guideline updates if a study is to be considered acceptable under USEPA's guideline (which has been updated much more recently [2012 versus 1984] than the OECD 206 guideline) and still be acceptable under the USEPA guideline. Therefore, if the current three dose group plus a control design is not sufficient to consistently generate robust BMDs, then additional adult animals must be added, so long as the total number of cages needed are still within the space constraints of the rooms in the contract laboratories conducting these studies. This is because dividing a study across multiple rooms is undesirable as it is another variable that must be accounted for in statistical analysis, potentially reducing power to detect an effect. Thus, seeking ways to reduce the number of offspring needed is a logical area to target to adhere to animal welfare concerns.

Working outside the constraints of current guidelines, there are many aspects of reproductive toxicity tests that could be altered to increase biological relevance (Mineau 2005). For example, OECD 206 guidelines use artificial incubation, no parental rearing, and artificial photostimulation, and the extended (8–10 week) egg-laying period means that date of clutch completion, a parameter that could be important for determining survival in the wild, is ignored as only total eggs laid is analyzed. Furthermore, chicks are only grown out to two weeks (in a caged environment with *ad libitum* food) and are then deemed to have "survived." It could be possible to increase biological relevance and generate data more suitable for regression-type approaches if the types of endpoints (adult health, eggshell effects, productivity, fertility, survival through incubation, hatchability, and survival of chicks) are separated out into different experiments that use advances in effects assessment (Rattner et al. 2023) at different levels of biological organization and include a combination of in vivo, in ovo, in vitro, and in silico tools.

Although updates to OECD guidelines are infrequent and take considerable time, there are small modifications that can be made to protocols that work within the existing guidelines. These small modifications can be implemented immediately in protocols and may help improve accuracy of data, e.g., standardizing chick stocking density in brooders to reduce variability in brooder survival rates or chick growth, cage modifications to improve quantification of food consumption and subsequently the dietary dose and gaining an understanding of the precision and accuracy with which eggshell thickness can be measured to aid interpretation of slight effects on this endpoint. Use of robust historical control datasets to evaluate areas within existing protocols that can be adjusted within the flexibility of the existing guidelines (without the need to invest time and resources for updates) to help improve accuracy and precision of data and aid study interpretation should be considered (Foudoulakis and Bean 2022).

Development of Toxicological Benchmarks Using Evidence Integration Techniques

Rarely do chemical-specific toxicity data publicly exist for the assessment endpoint species of interest. Laboratory wildlife models are limited, and species extrapolation methods and assumptions are often required. Additionally, assumptions regarding observation of effects within a controlled laboratory environment may differ markedly in importance compared with wildlife species in a multistressor environment. For example, a laboratory study that finds a 10% statistical reduction in egg production relative to treatment or dose may be insignificant when extrapolating that to wild populations where few nests escape predatory events. Ecologically, exposure to a substance that affects behavior (and subsequently territoriality or predator vigilance) may have a greater impact on reproductive success.

Several publications exist that provide recommendations to improve the collection and evaluation of data to develop toxicity reference values (TRVs; Allard et al. 2010; Deck and Johnson 2015; Mayfield et al. 2013). These methods primarily apply to contaminated sites where existing data are primarily used to develop TRVs. Since their publication, other methods developed for human health risk assessment have informed those used for TRV development. Examples include systematic review and evidence integration approaches.

Typically, a robust, documented literature search is the initial step where keywords, databases, dates, titles, abstracts, and relevant manuscripts are reported. The relevant toxicological literature should be summarized and each line of evidence evaluated to ascertain whether sufficient evidence exists for all observed toxic endpoints. Where sufficient evidence exists for a toxic effect (across all lines of evidence, i.e., in vitro, in vivo, read-across, field), then available dose–response information can be evaluated to determine a point of departure (POD) from which to develop a TRV . in vitro or mechanistic information can be useful in extrapolating effects to other species where it can be reasonably demonstrated that biological pathways are similar. Differences between species' physiologies can also provide evidence regarding potential differences in toxicokinetics. More accurate predictions can be made where there are species-specific physiologically based toxicokinetic models; however, these are often rare for many wildlife species. When these exist, in vitro/in vivo extrapolation can be useful in the development of PODs. It is important that *a priori* decisions not be made when extrapolating toxicity data for populational relevance since endpoints other than mortality, growth, and reproduction (e.g., lethargy in influencing predator avoidance or nest selection) can have profound indirect influence on metapopulational effects (Sample et al. 2022).

In some cases, tissue-based TRVs can be developed (e.g., blood lead; Johnson et al. 2004, 2007, 2020). Blood often can be obtained using minimally invasive means; however, other tissue samples (e.g., liver, kidney) require euthanasia. Tissue-based TRVs do significantly reduce much of the inherent uncertainty in exposure estimation.

Research and Data Needs

Clearly, many species-specific life-history information is lacking for many wildlife species (Morrissey et al. 2023; Sample et al. 2022). Allometric extrapolation based on body mass is still used primarily to estimate food ingestion rates. Collection of energetic requirements to include food choices are needed especially for many chiropteran species (bats), ruminants, birds, adult amphibians, and reptiles. Variation of energetics in ectodermic species is a challenge where oral exposures can be pulsed. Metabolism of xenobiotics is especially poorly understood in ectothermic species. Data for quantifying exposure from inadvertent soil or sediment ingestion, and dermal and inhalation exposures are lacking for many species of wildlife and in many cases cannot be estimated.

Wicked issues remain for understanding exposures to multiple substances in an exposome context (mixtures). Further complexity is added when variation of external stressors (e.g., climate change, food abundance and choices, and exposures within a heterogeneous environment) are considered. Many species are food limited, and reproductive output can be related to food availability (energetics). Toxicants that detract from energy balance through insult that require greater somatic maintenance may reduce reproductive output; however, this is based on theory and we are aware of no studies that have investigated this. Many populations are limited by predatory events, and factors that reduce those can vary between species and populations and have complex interactions (Creel and Christianson 2008; Cresswell 2011). Food abundance and predatory events may be proportional (McNamara and Houston 1987). Furthermore, relatively few examples exist where risk assessment conclusions are tested or verified from field studies. Dynamics of assessment populations (or metapopulations) may indeed be site and species specific where validating these relationships with field data is unlikely.

References

Allard, P., Fairbrother, A., Hope, B.K. et al. (2010). Recommendations for the development and application of wildlife toxicity reference values. *Integr. Environ. Assess. Manag.* 6: 28–37.

Aulerich, R.J., Ringer, R.K., and Iwamoto, S. (1974). Effects of dietary mercury on mink. *Arch. Environ. Contam. Toxicol.* 2: 43–51.

Baier, V., Paini, A., Schaller, S. et al. (2022). A generic avian physiologically based kinetic (PBK) model and its application in three bird species. *Environ. Int.* 169: 107547. https://doi.org/10.1016/j.envint.2022.107547.

Basu, N., Crump, D., Head, J. et al. (2019). EcoToxChip: a next-generation toxicogenomics tool for chemical prioritization and environmental management. *Environ. Toxicol. Chem.* 38: 279–288. https://doi.org/10.1002/etc.4309.

Bazar, M.A., Quinn, M.J. Jr., Mozzachio, K. et al. (2010). Toxicological responses of red-backed salamanders (*Plethodon cinereus*) exposed to aged and amended soils containing lead. *Arch. Environ. Contam. Toxicol.* 58: 1040–1047.

Bean, T.G., Arnold, K.E., Lane, J. et al. (2016). An *in vitro* method for determining the bioaccessibility of pharmaceuticals in wildlife. *Environ. Toxicol. Chem.* 35: 2349–2357. https://doi.org/10.1002/etc.3406.

Bean, T., Blakey, A., Ebeling, M. et al. (2022). Revision of the EFSA guidance document on birds and mammals: industry view. Platform Presentation 03.05.T-05.

Bean, T.G., Beasley, V.R., Berny, P. et al. (2023). Wildlife effects assessment in the 21st century: review of current effects assessment methods and recommendations for a path forward. *Integr. Environ. Assess. Manag.* https://setac.onlinelibrary.wiley.com/doi/10.1002/ieam.4795.

Bélisle, M. and Desrochers, A. (2002). Gap-crossing decisions by forest birds: an empirical basis for parameterizing spatially-explicit, individual-based models. *Landsc. Ecol.* 17: 219–231. https://doi.org/10.1023/A:1020260326889.

van den Berg, S.J.P., Maltby, L., Sinclair, T. et al. (2021). Cross-species extrapolation of chemical sensitivity. *Sci. Total Environ.* 753: 141800. https://doi.org/10.1016/j.scitotenv.2020.141800.

Beronius, A., Zilliacus, J., Hanberg, A. et al. (2020). Methodology for health risk assessment of combined exposures to multiple chemicals. *Food Chem. Toxicol.* 143: 111520. https://doi.org/10.1016/j.fct.2020.111520.

Bertram, M.G., Martin, J.M., McCallum, E.S. et al. (2022). Frontiers in quantifying wildlife behavioural responses to chemical pollution. *Biol. Rev.* 97: 1346–1364. https://doi.org/10.1111/brv.12844.

Beyer, W.N., Basta, N.T., Chaney, R.L. et al. (2016). Bioaccessibility tests accurately estimate bioavailability of lead to quail. *Environ. Toxicol. Chem.* 35 (9): 2311–2319. https://doi.org/10.1002/etc.3399.

Carlsen, T.M., Coty, J.D., and Kercher, J.R. (2009). The spatial extent of contaminants and the landscape scale: an analysis of the wildlife, conservation biology, and population modeling literature. *Environ. Toxicol. Chem.* 23 (3): 798–811. https://doi.org/10.1897/02-202.

Celander, M.C., Goldstone, J.V., Denslow, N.D. et al. (2011). Species extrapolation for the 21st century. *Environ. Toxicol. Chem.* 30: 52–63. https://doi.org/10.1002/etc.382.

Creel, S. and Christianson, D. (2008). Relationships between direct predation and risk effects. *Trends Ecol. Evol.* 23 (4): 194–201. https://doi.org/10.1016/j.tree.2007.12.004.

Cresswell, W. (2011). Predation in bird populations. *J. Ornithol.* 152 (Suppl 1): 251–263. https://doi.org/10.1007/s10336-010-0638-1.

Crump, D., Boulanger, E., Farhat, A. et al. (2020). Effects on apical outcomes of regulatory relevance of early-life stage exposure of double-crested cormorant embryos to 4 environmental chemicals. *Environ. Toxicol. Chem.* 40: 390–401.

Davidson, I.W.F., Parker, J.C., and Beliles, R.P. (1986). Biological basis for extrapolation across mammalian species. *Regul. Toxicol. Pharmacol.* 6 (3): 211–237. https://doi.org/10.1016/0273-2300(86)90014-0.

Deck, A.T. and Johnson, M.S. (2015). Methods for derivation of wildlife toxicity reference values for use in ecological risk assessments. In: *Wildlife Toxicity Assessments for Chemical of Military Concern* (ed. M.A. Williams, G. Reddy, M.J. Quinn Jr., and M.S. Johnson), 15–22. Amsterdam: Elsevier.

Dijak, W.D., Rittenhouse, C.D., Larson, M.A. et al. (2007). Landscape habitat suitability index software. *J. Wild. Manage.* 71 (2): 688–670. https://doi.org/10.2193/2006-341.

Dolman, P.M., Hinsley, S.A., Bellamy, P.E., and Watts, K. (2007). Woodland birds in patchy landscapes: the evidence base for strategic networks. *Ibis* 149 (2): 146–160. https://doi.org/10.1111/j.1474-919X.2007.00748.x.

Duggan, J.M., Eichelberger, B.A., Ma, S. et al. (2015). Informing management of rare species with an approach combining scenario modeling and spatially explicit risk assessment. *Ecosyst. Health Sustain.* 1 (6): 1–18. https://doi.org/10.1890/EHS14-0009.1.

Ebeling, M. and Wang, M. (2018). Dissipation of plant protection products from foliage. *Environ. Toxicol. Chem.* 37: 1926–1932.

Ebeling, M., Blakey, A., Foudoulakis, M. et al. (2023). Laying hen studies and time-to-effect in reproductive risk assessments for birds. May 2023, doi:10.13140/RG.2.2.13034.18880.

European Food Safety Authority (EFSA) (2022). Guidance on the use of the benchmark dose approach in risk assessment. *EFSA J.* 20 (10): e07584. https://doi.org/10.2903/j.efsa.2022.7584.

European Food Safety Authority (EFSA) (2023). Risk assessment for birds and mammals. *EFSA J.* 21 (2): e07790. https://doi.org/10.2903/j.efsa.2023.7790.

Farhat, A., Crum, D., Bidinosti, L. et al. (2019). An early–life stage alternative testing strategy for assessing the impacts of environmental chemicals in birds. *Environ. Toxicol. Chem.* 39: 141–154. https://doi.org/10.1002/etc.4582.

Fletcher, J., Nellesson, J.E., and Pfleeger, T.G. (1994). Literature review and evaluation of the EPA food-chain (Kenaga) nomogram, an instrument for estimating pesticide residues on plants. *Environ. Toxicol. Chem.* 13 (9): 1383–1391.

Foudoulakis, M. and Bean, T. (2022). Avian reproductive toxicity studies – ecotoxicologist perspective. EFSA preparatory work on how to report, use and interpret historical control data in (eco)toxicity studies. May-3-5 2022, online, *in* Tamara Coja, Agathi Charistou, Aikaterini Kyriakopoulou, Kyriaki Machera, Ulrike Mayerhofer, Dimitra Nikolopoulou, Eliana Spilioti, Anastasia Spyropoulou, Johann Steinwider, Tanja Tripolt (2022) Preparatory work on how to report, use and interpret historical control data in (eco)toxicity studies. *EFSA Support. Publ.* 19 (9): 7558E. https://doi.org/10.2903/sp.efsa.2022.EN-7558.

Galic, N., Ashauer, R., Gergs, A. et al. (2022). What mechanistic effect models can and cannot tell us about the variability in the real world. EFSA ONE conference 21–24 June 2022, Brussels, Belgium and Online.

Gibbons, D., Morrissey, C., and Mineau, P. (2015). A review of the direct and indirect effects of neonicotinoids and fipronil on vertebrate wildlife. *Environ. Sci. Pollut. Res.* 22: 103–118. https://doi.org/10.1007/s11356-014-3180-5.

Gobas, F.A.P.C., Burkhard, L.P., Doucette, W.J. et al. (2016). Review of existing terrestrial bioaccumulation models and terrestrial bioaccumulation modelling needs for organic chemicals. *Integr. Environ. Assess. Manag.* 12: 123–134.

Grasman, K.A. (2002). Assessing immunological function in toxicological studies of avian wildlife. *Integr. Comp. Biol.* 42 (1): 34–42. https://doi.org/10.1093/icb/42.1.34.

Green, J.W. et al. (2022). https://doi.org/10.1186/s12302-022-00603-5.

Green, J.W., Foudoulakis, M., Bean, T. et al. (2023). Avian reproduction studies: experimental design and statistical methodology. Poster Presentation, SETAC Europe 33rd Annual Meeting, Dublin.

Harwood, J. (2000). Risk assessment and decision analysis in conservation. *Biol. Conserv.* 95: 219–226. https://doi.org/10.1016/S0006-3207(00)00036-7.

Heinz, G.H., Hoffman, D.J., Klimstra, J.D. et al. (2009). Species differences in the sensitivity of avian embryos to methylmercury. *Arch. Environ. Contam. Toxicol.* 56: 129–138. https://doi.org/10.1007/s00244-008-9160-3.

Herberstein, M.E., D.J., M.L., Lowe, E. et al. (2022). Animal Traits – a curated animal trait database for body mass, metabolic rate and brain size. *Sci. Data* 9 (1): 265. https://doi.org/10.1038/s41597-022-01364-9.

Hickie, B.E., Ross, P.S., MacDonald, R.W., and Ford, J.K.B. (2007). Killer whales (*Orcinus orca*) face protracted health risks associated with lifetime exposure to PCBs. *Environ. Sci. Technol.* 41: 6613–6619. https://pubs.acs.org/doi/10.1021/es0702519.

Hobday, A.J., SmithADM, S.I.C., Daley, R. et al. (2011). Ecological risk assessment for the effects of fishing. *Fish. Res.* 108: 372–384. https://doi.org/10.1016/j.fishres.2011.01.013.

Hoerger, F. and Kenaga, E.E. (1972). Pesticide residues on plants: correlation of representative data as a basis for estimation of their magnitude in the environment. In: *Environmental Quality and Safety: Chemistry, Toxicology and Technology*, vol. 1 (ed. F. Coulston and F. Corte), 9–28. Stuttgart: Georg Theime Publishers.

Hope, B.K. (2000). Generating probabilistic spatially-explicit individual and population exposure estimates for ecological risk assessments. *Risk Anal.* 20: 573–590. https://doi.org/10.1111/0272-4332.205053.

Hope, B.K. (2002). A case study comparing static and spatially explicit ecological exposure analysis methods. *Risk Anal.* 21 (6): 1001–1010. https://doi.org/10.1111/0272-4332.216169.

Hope, B.K. (2004). Approaches to spatially-explicit, multi-stressor ecological exposure estimation. In: *Landscape Ecology and Wildlife Habitat Evaluation* (ed. L. Kapustka, G. Biddinger, M. Luxon, and H. Galbraith), 311–324. Conshohocken, PA: STP 1458, ASTM International.

Hussain, I., Khan, M.Z., Khan, A. et al. (2008). Toxicological effects of diclofenac in four avian species. *Avian Pathol.* 37 (3): 315–321. https://doi.org/10.1080/03079450802056439.

Jager, T., Vandenbrouck, T., Baas, J. et al. (2010). A biology-based approach for mixture toxicity of multiple endpoints over the life cycle. *Ecotoxicology* 19 (2): 351–361. https://doi.org/10.1007/s10646-009-0417-z.

Johnson, M.S. and McAtee, M.J. (2015). Characterizing potential for toxicity: estimating risks to wildlife. In: *Wildlife Toxicity Assessments for Chemical of Military Concern* (ed. M.A. Williams, G. Reddy, M.J. Quinn Jr., and M.S. Johnson), 1–13. Amsterdam: Elsevier.

Johnson, M.S. and Pillard, D. (2023). Considerations of amphibian models and testing paradigms for ecological risk assessment in non-conventional vertebrate species in environmental risk assessment strategies. In: *Issues in Toxicology*, vol. 3 (ed. M.L. Larramendy and G.E. Liwszyc). Cambridge: Royal Society of Chemistry. (in press).

Johnson, M.S., Major, M.A., and Casteel, S.W. (2004). Lead accumulation in woodchucks (*Marmota monax*) at small arms and skeet ranges. *Ecotoxicol. Environ. Saf.* 59 (2): 232–236.

Johnson, M.S., Wickwire, W.T., Quinn, M.J. et al. (2007). Are songbirds at risk from lead at small arms ranges? An application of the spatially explicit exposure model (SEEM). *Environ. Toxicol. Chem.* 26: 2215–2225.

Johnson, M.S., Korcz, M., von Stackelberg, K.L., and Hope, B.K. (2009). Spatial analytical techniques for risk based decision support systems. In: *Decision Support Systems for Risk-Based Management of Contaminated Sites* (ed. A. Marcomini, G.W. Suter II, and A. Critto), 75–93. New York, NY: Springer.

Johnson, M.S., McFarland, C.A., Bazar, M.A. et al. (2010). Toxicity of octahydro-1,3,5,7-tetranitro-1,3,5,7-tetrazocine (HMX) in three vertebrate species. *Arch. Contam. Toxicol.* 58: 836–843.

Johnson, M.S., Quinn, M.J., Wickwire, T. et al. (2020). Do spatially explicit wildlife exposure models improve the estimation of risk for small mammals? Case study: application of SEEM to small mammal exposures to lead in heterogeneous landscapes. *Integr. Environ. Assess. Mngt.* 17: 259–272. https://doi.org/10.1002/ieam.4326.

Kuo, D.T.F., Rattner, B.A., Marteinson, S.C. et al. (2022). A critical review of bioaccumulation and biotransformation of organic chemicals in birds. *Rev. Environ. Contam. Toxicol.* 260: 6. https://doi.org/10.1007/s44169-021-00007-1.

Lahr, J., Art, G., Duquesne, S. et al. (2023). Proposal of critical appraisal tools for the evaluation of ecotoxicology studies. *EFSA Support. Publ.* 20 (3): 7787E. https://doi.org/10.2903/sp.efsa.2023.EN-7787.

Lee, J.M., Lee, B.H., Chang, S.N. et al. (2018). Establishment, characterization, and toxicological application of a spontaneous immortalized cell line from the striped field mouse, *Apodemus agrarius. in vitro Cell. Dev. Biol.-Animal* 54: 779–787. https://doi.org/10.1007/s11626-018-0290-2.

Ludwigs, J.D., Ebeling, M., Fredricks, T.B. et al. (2017). Appropriate exposure estimates for wildlife risk assessments of crop protection products based on continuous radio telemetry: a case study with wood pigeons. *Environ. Toxicol. Chem.* 36: 1270–1277.

Ludwigs, J.D., Ebeling, M., Haaf, S., and Kragten, S. (2022). Monte Carlo PT simulations for wildlife risk assessment: assessing the portion of diet taken by birds and mammals from a pesticide-treated area – proposal for a joint way forward. *Environ. Toxicol. Chem.* 41: 1344–1354. https://doi.org/10.1002/etc.5311.

Matson, C.W., Gillespie, A.M., McCarthy, C. et al. (2009). Wildlife toxicology: biomarkers of genotoxic exposures at a hazardous waste site. *Ecotoxicology* 18: 886–898. https://doi.org/10.1007/s10646-009-0350-1.

Mayfield, D.B., Johnson, M.S., Burris, J.A., and Fairbrother, A. (2013). Furthering the derivation of predictive wildlife toxicity reference values for use in soil cleanup decisions. *Integr. Environ. Assess. Manag.* 10 (3): 358–371.

McNamara, J.M. and Houston, A.I. (1987). Starvation and predation as factors limiting population size. *Ecology* 68: 1515–1519. https://doi.org/10.2307/1939235.

Mineau, P. (2005). A review and analysis of study endpoints relevant to the assessment of "long term" pesticide toxicity in avian and mammalian wildlife. *Ecotoxicology* 14: 775–799.

Mitra, A., Chatterjee, S., Sarkar, M., and Gupta, D.K. (2021). Toxic effects of pesticides on avian fauna. In: *Environmental Biotechnology Vol. 3. Environmental Chemistry for a Sustainable World*, vol. 50 (ed. K.M. Gothandam, S. Ranjan, N. Dasgupta, and E. Lichtfouse). Cham: Springer. https://doi.org/10.1007/978-3-030-48973-1_3.

Morrissey, C., Fritsch, C., Fremlin, K. et al. (2023). Advancing exposure assessment approaches to improve wildlife risk assessment. *Integr. Environ. Assess. Manag.* https://doi.org/10.1002/ieam.4743.

Organization for Economic Co-operation and Development (OECD) (1984). OECD 206: avian reproduction test, OECD guideline for the testing of chemicals. https://www.oecd-ilibrary.org/environment/revised-guidance-document-150-on-standardised-test-guidelines-for-evaluating-chemicals-for-endocrine-disruption/avian-reproduction-test-oecd-tg-206_9789264304741-16-en#:~:text=TG%20206%20is%20an%20OECD%20validated%20test%20to,dietary%20concentrations%20of%20the%20test%20chemical%20are%20required.

Peakall, D.B. (1970). Pesticides and the re-production of birds. *Sci. Am.* 222: 73–78.

Purucker, S.T., Welsh, C.J.E., Stewart, R.N., and Starzec, P. (2007). Use of habitat-contamination spatial correlation to determine when to perform a spatially explicit ecological risk assessment. *Ecol. Model.* 204 (1–2): 180–192. https://doi.org/10.1016/j.ecolmodel.2006.12.032.

Raimondo, S., Lilavois, C.R., and Barron, M.G. (2015). Web-based interspecies correlation estimation (web-ICE) for acute toxicity: user manual. Version 3.3 (EPA/600/R-15/192). U. S. Environmental Protection Agency, Office of Research and Development, Gulf Ecology Division.

Rattner, B.A. (2009). History of wildlife toxicology. *Ecotoxicology* 18: 773–783.

Rattner, B.A., Bean, T.G., Beasley, V.R. et al. (2023). Wildlife ecological risk assessment in the 21st century: promising technologies to assess toxicological effects. *Integr. Environ. Assess. Manag.* https://doi.org/10.1002/ieam.4806.

Reed, J.M. (2001). The role of behavior in recent avian extinctions and endangerments. *Conserv. Biol.* 13: 232–241. https://doi.org/10.1046/j.1523-1739.1999.013002232.x.

Romanov, M.N., Tuttle, E.M., Houck, M.L. et al. (2009). The value of avian genomics to the conservation of wildlife. *BMC Genomics* 10 (Suppl 2): S10. https://doi.org/10.1186/1471-2164-10-S2-S10.

Saaristo, M., Brodin, T., Balshine, S. et al. (2018). Direct and indirect effects of chemical contaminants on the behaviour, ecology and evolution of wildlife. *Proc. R. Soc.* B285: 20181297. http://dx.doi.org/10.1098/rspb.

Sample, B.E., Johnson, M.S., Hull, R.N. et al. (2022). Key challenges and developments in wildlife ecological risk assessment: problem formulation. *Integr. Environ. Assess. Manag.* https://doi.org/10.1002/ieam.4710.

Scanes, C.G., Witt, J., Ebeling, M. et al. (2022). Quantitative comparison of avian and mammalian physiologies for parameterization of physiologically based kinetic models. *Front. Physiol.* 5 (13): 858386. doi: 10.3389/fphys.2022.858386.

Sih, A., Cote, J., Evans, M. et al. (2012). Ecological implications of behavioural syndromes. *Ecol. Lett.* 15: 278–289. https://doi.org/10.1111/j.1461-0248.2011.01731.x.

Sorensen, M.T., Gala, W.R., and Margolin, J.A. (2004). Approaches to ecological risk characterization and management: selecting the right tools for the job. *Hum. Ecol. Risk Assess. Int. J.* 10 (2): 245–269. https://doi.org/10.1080/10807030490438193.

von Stackelberg, K., Williams, M.A., Clough, J., and Johnson, M.S. (2017). Spatially-explicit bioaccumulation modeling in aquatic environments: results from two demonstration sites. *Integr. Environ. Assess. Manag.* 13 (6): 1023–1037.

Tannenbaum, L.V., Johnson, M.S., and Bazar, M. (2003). Application of the hazard quotient method in remedial decisions: a comparison of human and ecological risk assessments. *Hum. Ecol. Risk Assess. Int. J.* 9 (1): 387–401. https://doi.org/10.1080/713609871.

Tobias, J.A., Sheard, C., Pigot, A.L. et al. (2022). AVONET: morphological, ecological and geographical data for all birds. *Ecol. Lett.* 25 (3): 581–597. https://doi.org/10.1111/ele.13898.

Trijau et al. 2023 is https://doi.org/10.1016/j.envpol.2023.121477.

US EPA 1993 is the wildlife exposure factors handbook, available here. https://assessments.epa.gov/risk/document/&deid=2799.

UBA (Umwelt Bundesamt, German Environment Agency) (2023). Review of the OECD test guidelines relevant to environmental assessment with regard to the state of the art in science and technology – Final report. http://www.umweltbundesamt.de/publikationen.

United States Environmental Protection Agency (USEPA) (2012a). T-REX version 1.5 user's guide for calculating pesticide residues on avian and mammalian food items user's guide T-REX version 1.5 (terrestrial residue EXposure model). https://www.epa.gov/pesticide-science-and-assessing-pesticide-risks/t-rex-version-15-users-guide-calculating-0https://www.epa.gov/pesticide-science-and-assessing-pesticide-risks/t-rex-version-15-users-guide-calculating-pesticide#:~:text=T-REX%20calculates%20the%20residues%20on%20avian%20and%20mammalian,model%20also%20estimates%20acute%20and%20reproductive%20risk%20quotients

United States Environmental Protection Agency (USEPA) (2012b). OCSPP 850.2300: avian reproduction test, ecological effects test guidelines. https://www.regulations.gov/document/EPA-HQ-OPPT-2009-0154-0012.

Verboom, J., Foppen, R., Chardon, P. et al. (2001). Introducing the key patch approach for habitat networks with persistent populations: an example for marshland birds. *Biol. Conserv.* 100 (1): 89–101. https://doi.org/10.1016/S0006-3207(00)00210-X.

9

Reflections from an Environmental Toxicologist Working in a Large Company

Ralph G. Stahl Jr (Retired)

DuPont Company, Corporate Environmental Affairs and Environmental Toxicology, Wilmington, DE, USA

Introduction

Why would anybody want to work as an environmental toxicologist in a large company? Wouldn't you be able to do "better" by working for a governmental environmental agency or a nonprofit conservation organization? Both are good questions, and in this chapter hopefully, I will answer them based on my 34-year career as an environmental toxicologist at the DuPont Company. I've chosen to write this chapter in a less technical format than one might expect – more from a personal perspective and including details about my experience that I have not written about or published previously.

To help with my response to the questions, consider what I view as the three main scientific trends I experienced during my career: (1) the beginning and subsequent widespread use of ecological risk assessment and risk management; (2) climate change and its influence on the practice of environmental toxicology and chemistry; and (3) remediation and restoration of contaminated land/habitats. In effect, all three of these were focal points in what I was tasked with for most of my career, perhaps none more so than remediation and restoration of contaminated land. And, in a foreshadowing of the discussion section, these trends were the stimuli for joining the Society of Environmental Toxicology and Chemistry [SETAC] and for participating in numerous SETAC meetings and workshops over the years.

I will also touch on the need to learn and develop people skills, which are seldom taught in graduate school or during a postdoctoral fellowship. If you retain anything from this chapter, let it be that learning and continuing to improve your people skills will serve you well in your career and elsewhere. Strong technical competence is important, but it will not help you to address as many challenges as you would be able to when you can work and communicate well with other people.

Remediation and Restoration of Contaminated Land in the United States

Remediation of contaminated current and former manufacturing sites (land) is a major effort for many large companies in the United States. Much of this work is conducted under two federal statutes, the Comprehensive Environmental Response Compensation and Liability Act [or Superfund] (U.S. Congress 1980) and the Resource Conservation and Recovery Act (RCRA, U.S. Congress 1976), as well as state statutes. Some 1,330 Superfund sites are listed by the U.S. Environmental Protection Agency (USEPA) on the National Priorities List (NPL) (https://www.epa.gov/superfund/superfund-national-priorities-list-npl), while approximately 3,600 sites are on the RCRA Corrective Action (RCRA-CA) list (https://www.epa.gov/hwcorrectiveactioncleanups/corrective-action-programs-around-nation). These statutes have been in place for decades, and the state of practice for civil engineers, hydrologists, geologists, chemists, toxicologists, and risk assessors and the interactions between companies and regulatory agencies, etc. are well developed.

To deal with the remediation obligations under these two statutes, some companies maintain internal technical and legal staff to manage the remediation program under a corporate umbrella. They can represent sizable elements of cost, or a focus of public concern, and therefore need to be managed well. In some cases, the annual costs in a company can exceed US$10 million. The details on the known or projected costs for meeting remediation and other environmental obligations can be found in a company's annual reports or in their annual Securities and Exchange Commission Form 10-K (US SEC 2020).

The publicly available financial information, coupled with the number of sites across the nation that are in or yet to be in the Superfund or RCRA program, indicates clearly that remediation of current or future manufacturing sites will be ongoing for many years to come. In turn, this suggests that corporate remediation programs and their associated internal staff will also be needed into the future, or perhaps the management and work involved will be contracted to a third party to reduce corporate staff and costs.

Remediation and Restoration of Contaminated Land Outside the United States

In contrast to the United States, the legal and regulatory frameworks and the state of practice for remediating contaminated sites elsewhere is diverse, less well developed, and thus not as well understood by those who are tasked with addressing contamination at a particular site. In some countries, it is difficult to know how many contaminated sites exist, what experiences other companies might have gained by remediating and restoring them, and, most importantly, whether the equipment needed for investigations, chemical analysis, or implementing a remedial action is readily available. It also is not clear whether an existing manufacturing site, or a formerly owned site that has been divested, might be the subject of regulatory action.

It is not surprising that practical experience in environmental toxicology or risk assessment gained at US sites may not translate well into the needs in other countries. In addition to potential language barriers, sometimes a lack of understanding of cultural norms and how the regulatory process works in a specific country are issues.

To complicate matters even further, remedial programs within a company are also challenged when the site is outside the United States. For example, it might be a joint venture with one or more non-US company(s), or is owned and operated by an in-country subsidiary of the parent company (Kovalick and Montgomery 2017), or, as in some countries, be owned in part by a State entity. It is often the case that these complicated situations boil down to questions of which company (parent or subsidiary) is responsible for the remediation, who will do the technical and legal work associated with remediation, as well as how and where to find reputable contractors in-country that have the equipment and expertise to implement a remedial project in a safe and effective manner. From the outside, it may seem to be an exciting career scenario for an environmental toxicologist, traveling and working in other countries, meeting lots of new people, etc. It can be exciting, but that is not true in every case.

What Is a "Site"

During my career, I was involved with remediation of numerous sites inside and outside the United States. To begin with, it is important to recognize that regardless of what country a "site" is located, seldom it is where a single product is manufactured – in most cases multiple products are produced, often by different business subunits operating under the larger umbrella of a parent company. It is not unusual for the site to be owned by one subunit of the company yet occupied by several other subunits that act as "tenets." As tenets the subunits use all the site facilities and pay rent as they do not own the site per se. The owner/operator business subunit then is responsible for operating and maintaining the infrastructure, providing site security and safety, and managing interactions with regulatory agencies to obtain and comply with the various permits required for operations.

The diversity within a site's operations is not obvious but does provide an interesting dynamic when a site is required to deal with remediation, whether in the United States under Superfund or RCRA, or a state regulatory program, or under the various statutes in other countries. In the United States, for example, one can ask whether the contaminant that poses the most risk to human health and the environment is coming from one specific production line, and, if so, shouldn't that business subunit producing it pay for the remediation? This issue also plays out most frequently at multiparty Superfund sites where companies typically resort to hiring a third-party allocator whose job is to apportion a share of remedial costs based on which contaminants are the basis for most or all of the remediation. To avoid these contentious issues at multiparty Superfund sites, often an initial agreement is made by the companies to address the regulatory requirements on a per-capita basis. Even so, there is an underlying knowledge that at some point the costs will likely be allocated as some companies may feel dissatisfied paying for remedial work they do not feel pertains to their operations or products.

DuPont Remediation Program

Before 1990, responsibility for dealing with remedial issues at DuPont was at the site or business subunit level. In other words, each site and the business subunit that owned or operated it managed the remediation themselves. DuPont had been in operation since 1802, with more than 100 sites across the world and numerous business subunits that manufactured and sold any number of products carrying the DuPont Company label. As one can imagine, with decentralized responsibility and accountability, there were wide variations in how remediation and restoration was addressed. This decentralization also resulted in a loss of efficiency as no national or corporate-wide contracts for specific remediation services (constructing groundwater monitoring wells, chemical analyses of environmental media, and public communications, among others) were in place.

With the advent of a centralized remediation group and associated staff in 1990, national corporate contracts for various aspects of remedial design and construction, and corporate expertise, could be brought to bear and help reduce or eliminate excess transactional costs. It also provided a platform for internal staff to share key learnings and to fund research on specific topics relevant to a remediation issue. Interestingly, during my tenure, some of those funds were directed to specific topics that were addressed through the SETAC Pellston© Workshop program.

The DuPont CRG was composed of roughly 30–40 individuals based in the United States, Europe, Asia/Pacific, and Latin America. It included a variety of senior technical experts, many of whom had years of experience working in the field of remediation. Where possible, we applied risk-based decision-making at each site to help determine how best to manage risks to human health and the environment. As one might expect, we learned early on that our decision-making process would need to accommodate the suite of stressors we faced at the various sites.

Risk-Based Decision-Making

For the purposes of this discussion, risk-based decision-making is defined as a three-step process: (1) collect and evaluate environmental data to estimate potential exposures to humans and ecological receptors; (2) estimate risks to humans and ecological receptors from those exposures; and (3) incorporate the results into future land use plans for developing a list of potential remedial risk management options. Some of these steps and the risk-based decisional process have been discussed in a broader context (Kapustka et al. 2016).

At the same time, the decisional process required that the cost of the work is factored in, among other considerations including regulatory requirements, the business subunit responsible for owning and operating the site, other business subunits also using the site, company policies, and local community interests. How sites operate and companies address remedial issues are not unique to DuPont, as I learned from my colleagues at other companies with responsibilities like mine.

I discuss three site/case examples to illustrate the points about managing multiple stressors while remediating and restoring contaminated land. Two of the sites are in the United States and have been undergoing or have finished remediation under either Superfund or RCRA statutes. I have written about these two sites previously, but not with some of the personal insights contained herein. The third site is in the United Kingdom and was selected to illustrate how the remediation program within a company can differ from that which would apply if the site were in the United States. I have not written about this latter one. It is also an example of how the legal and regulatory requirements for remediation and restoration of contaminated land differ from those in the United States.

Sites/Examples

The three sites/examples are a small subset of the many sites that had been under the purview of the DuPont Corporate Remediation Group (CRG). After my retirement, most of the DuPont CRG was split between the "new" DuPont and Corteva Agrisciences due to a merger and divestiture of various business units. Today, the approach to remediation and restoration may differ from what I experienced and describe here.

The United States – Superfund Site, Newport, Delaware

The DuPont Newport, Delaware Superfund site was one of the first to be evaluated in a risk-based decisional manner under the DuPont CRG in early 1990, and some details have been published previously (Stahl et al. 2001). It is unique in that the focus of the remedy was to protect ecological receptors rather than humans alone. This was a departure from typical practice and novel for the regulatory agencies. Fortunately, work benefitted from the early draft or published USEPA guidance documents on ecological risk assessment (USEPA 1992, 1995).

Briefly, some soil and sediment in wetlands at the Newport site and in the nearby Christina River had been contaminated with metals. The environmental data were collected and evaluated by a site remediation team that included the USEPA, DuPont, and the State of Delaware. In parallel, and under the Superfund framework, the natural resource Trustees (US Fish and Wildlife Service (USFWS), the National Oceanic and Atmospheric Administration (NOAA), and the State of Delaware) participated, providing their input with respect to protecting and restoring the resources under their purview (Stahl et al. 2001). The interactions among this group resulted in a USEPA Record of Decision (ROD) (USEPA 1993) in 1993 that required removal of soils and sediments in affected areas, backfilling with clean soil, and replanting vegetation. While the ROD suggested one set of actions and the costs associated with them were known, it was also recognized that there could be a need (potentially but not yet quantified) to satisfy the natural resource Trustee's desire for restoration should a natural resource damage claim be filed (USDOI 2017).

The need to satisfy a restoration requirement led to another first in the CRG: an analysis was conducted to determine the benefits and costs associated with coupling remediation with restoration. Could remediation and restoration be coordinated in such a way as to remove a greater mass of contaminated soil and sediment, while providing a much more diverse habitat in the process? The analysis was based on projections of potential habitat value increases if restoration were coordinated with the remediation, judged against what could be achieved with the ROD alone. This is not unlike the process used by natural resource Trustees today, where the potential "uplift" in natural resource services for a particular action is estimated and serves as the basis for restoration settlement discussions (Dunford et al. 2004). At Newport, the results suggested that there would be little or no extra cost involved with coordinating the two efforts, and greater ecological services could be achieved. Over time, postremedial monitoring showed the projections to be accurate.

It is important to mention as well that other modifications to the remedy were considered but in areas where coordination of remediation and restoration would not be feasible. One area was an old landfill where the cover was not replanted but instead made to support a solar panel array.

A discussion of the Newport site would not be complete without noting how a potential natural resource damage claim was resolved and how the solution influenced our approach to remediation and restoration at other sites in the United States.

Even though the coordination of remediation and restoration did provide ecological service gains to help offset a potential damage claim, they were not sufficient to offset it completely. A small team comprising DuPont, USFWS, NOAA, and the State of Delaware met and determined how much additional restoration and the subsequent ecological service gains would be needed to satisfy the claim. Ultimately, more than 30 possible options were evaluated, the majority of which were parcels of land in the Christina River watershed that could be purchased outright and donated to the State of Delaware, or lands in the watershed that needed some restoration action to improve habitat quality. Unfortunately, none of these options proved feasible for one reason or another. Instead, an option was selected to purchase, restore, and donate land outside the Christina River watershed, which provided sufficient ecological service gains to satisfy the restoration requirement.

The significant work that went into identifying and securing the appropriate option became the basis for the concept "restoration banking," later changed to "restoration up front" (Stahl et al. 2008). The publication was in response to the time and effort that went into finding a suitable restoration option in an area where numerous potential options were found but were not suitable for various reasons. Our collective view was that it would be useful for companies and regulatory agencies to have a process or policy in place that would afford a mechanism to obtain suitable restoration options for potential use in the future or to transfer those options to another entity should they be needed to satisfy a claim. Today, such policies exist within two of the major federal natural resource Trustee agencies – NOAA and the US Department of the Interior (Taylor 2007; USDOI 2021).

Much of what we learned at this site became part of the corporate process we applied at other sites where we would be working with a diverse set of regulatory requirements and staff, as well as where there might be a potential for a claim for natural resource damages.

The United States – RCRA-Corrective Action Site, Waynesboro, Virginia

The Waynesboro, Virginia site, and the adjacent South River and associated watershed has been the subject of numerous publications (www.southriverwatershed.org). In contrast to the Newport, Delaware site where multiple metal contaminants were the target of remedial action, at Waynesboro and the South River, a single contaminant, mercury, was found in the environment and was the focus of numerous investigations and ultimately remediation.

Like the Newport, Delaware site, the work at Waynesboro and the South River also provided key insights into managing corporate remediation and working with regulatory agencies. From the start of our interactions with regulators, there was shared interest to understand why previous projections that mercury in the South River and its biota would attenuate naturally did not prove accurate (Stahl 2022). After several meetings, our shared goals became the basis for forming a multistakeholder group named the "South River Science Team." This team became the oversight group through which we planned, designed, and implemented investigations and remedial actions, as well as monitored the effectiveness of those actions. The South River Science Team was formed in late 2001 and operated under that name until 2022 when it was renamed the South River Watershed Coalition (see www.southriverwatershed.org).

The second feature of the Waynesboro and South River site was the unusual dynamic that resulted from having multiple agencies and organizations involved with it, each having a slightly different set of concerns. First, the remedial investigations and subsequent requirements for remedial and restoration work were under RCRA and overseen by the Virginia Department of Environmental Quality (VDEQ), and Region 3 of the USEPA. Second, the USFWS and the Commonwealth of Virginia were interested in pursuing data collection and analyses as part of a potential natural resource damage claim. Third, there was the potential for civil litigation under the citizen suit provision in RCRA, which in this case was being considered by the Natural Resources Defense Council and the Virginia Chapter of the Sierra Club.

As I foreshadowed in the Introduction, how to work and communicate with people is not taught in college but often gained through experience – sometimes the hard way. In my case, the Waynesboro and South River site was my training ground, with bumps along the way.

The need for developing people skills was driven by having three main "drivers" in our work on the Waynesboro site and nearby South River. Dealing with the requirements under RCRA that involved the USEPA and the Commonwealth of Virginia was perhaps the least stressful – we could follow regulations and work cooperatively with regulatory staff since we shared a common goal. The same was true for the potential natural resource damage claim – the staff from the USFWS and the Commonwealth of Virginia wanted to collect data, analyze it, and develop restoration options in a cooperative manner, without resorting to litigation. In fact, a cooperative assessment agreement among us had been signed early in our interactions, allowing us to work cooperatively within a legal construct and avoid conflicts on data collection, analyses, and evaluating restoration options. We also learned the benefit of exploring restoration options early, so we could avoid the complications we faced at the Newport site.

The more difficult of the situations was how to address the potential for civil litigation from the Natural Resources Defense Council and Virginia Chapter of the Sierra Club. None of us on the DuPont team had experience working with the two groups, nor had we been faced with potential civil litigation under the RCRA statute. It was made more difficult as well because there were no regulatory guidelines in place to keep us aligned on how to address the mercury problem. On the one hand, we were already working under RCRA, with oversight from the USEPA and the Commonwealth of Virginia. Yet it was not clear how that meshed, or not, with the potential civil action. Fortunately, as our interactions with the Natural Resources Defense Council, their outside academic expert, and others evolved, we gained a better insight as to what was driving their reliance on potential civil litigation. Even with the "bumps in the road," we were able to work closely with them to reach agreement on additional investigations and monitoring, as well as other public health actions they considered important.

Last, the Waynesboro and South River site was the first time we were asked to consider the potential impacts of climate change as we developed remedial options for addressing the mercury contamination. This consideration manifested itself in several ways. For example, in streambanks where mercury still resided, we did not uproot old-growth trees or conduct wholesale excavation of contaminated bank soils and sediments. Loss of old-growth trees and their ability to offset surface water temperature increases during drier, summer months would be very harmful to the ecological receptors inhabiting the areas. Instead, shallow excavation around these old-growth trees, coupled with backfill of clean soils (also augmented with soil amendments [biochar] to bind residues of mercury), and revegetation, was conducted. Insights into how to address the potential for changes if the climate warmed stemmed from several studies conducted over multiple years (Landis et al. 2017).

The United Kingdom – Contaminated Lands Program Site, Northern Ireland

The site in Northern Ireland was one of the first ones where remediation was being conducted under the United Kingdom's Model Procedures for the Management of Contaminated Land (DEFRA 2004, 2006). The site is located about 20 km outside of Londonderry, abutting the Faughan River, and had been owned and operated before DuPont purchased it in the 1950s. It also served as a landing zone for Allied military aircraft during the Second World War.

The contaminants of concern include volatile organic substances in groundwater that were detected periodically in river surface water, as well as large impoundments of spent lime from the previous owner's operations, as well as some metals found in site soils. The site is also noteworthy in that the part of it that abuts the Faughan River is designated as an ASSI – area of special scientific interest. This designation was due to its proximity to the Faughan River, which has a seasonal run of wild Atlantic salmon. Due to the presence of wild salmon, the Faughan River upstream of the DuPont site is designated as a Special Area of Conservation or SAC (https://www.daera-ni.gov.uk/protected-areas/river-faughan-and-tributaries-sac).

Because the Faughan River runs along the site and underneath a highway bridge, the infrastructure on either end of the bridge offers a stable and easily accessible location for local anglers, especially active during the time when salmon fishing is allowed. Another feature of the site is how the local community is involved with protection and conservation of the Faughan River. A local fisherman's association has a remit to keep the Faughan River along the site and its upstream reaches in good shape, including contacting local enforcement officials if individuals are seen attempting to catch salmon illegally. This fisherman's association became a good sounding board for us when we needed to gain some insight on how the local community viewed our remedial plans.

Despite its location, this site was not unlike other DuPont sites I had worked on in the United States. I found that quite a few of the DuPont sites are located on or near rivers or coastal areas. Proximity to water bodies provided not only a source of water for site operations but also means of transportation of finished products. The downside, though, was having to deal with contaminants in soils as well as in surface water and sediments.

The site in Northern Ireland gave me insight into how my experience in the United States did not translate completely. In this case, the remediation and other site-related regulations are managed by the Northern Ireland Environment Agency (NIEA). For remediation, their regulatory framework is the United Kingdom's Model Procedures for the Management of Contaminated Land, which differs from Superfund or RCRA. For example, in the United States, we received direct, and sometimes hands-on, guidance on investigations and other aspects of site work from regulators, which was not the case with NIEA. The staff at NIEA would review and provide comments on proposed work, even specific suggestions sometimes, but seldom offer specific requirements or impose mandates. NIEA staff, like the staff at USEPA, were well trained and highly competent but saw their role differently as it pertains to regulatory compliance and oversight for site remediation. This approach would be welcome by some who are accustomed to having fairly detailed guidelines and policies (sometimes called overly prescriptive), as well as direct requests and requirements from US regulatory staff. The downside is when there is considerable time spent drafting, submitting, and then redrafting proposals for investigations or remediation. It might make more sense to have a specific method to follow and comply with yet more modifications based on site-specific conditions. Fortunately, work at this site went well as the NIEA staff were willing to participate in multiple meetings, project reviews, and conference calls to help us manage the remedial needs in a timely manner.

Another feature of the Northern Ireland site and others in DuPont is that site land is split into two distinct sections: one that contains the infrastructure needed to manufacture and ship the finished product(s), and the other section that is left in "as is" condition since the time the land was purchased. In other words, these "as is" sections could be quite large, not subject to site operations directly, and contain habitats that had not changed significantly over many years. The on-site area adjacent to the Faughan River fits this category and is designated an ASSI.

Having a section of a site in "as in" condition, as was true for the Newport, Delaware site, and in Northern Ireland, can be beneficial. Some portions of the land might be valuable habitats, which can be conserved or restored to help satisfy a regulatory need. The quality of the habitat in the "as is" section could make it suitable for donation to a local or national conservation organization. Clearly though, an investigation into potential contamination of soils and other environmental media in that section of the site would be needed before any such actions could be undertaken. Perhaps, in another situation, the "as is" section may prove suitable for coordinating remediation with restoration, should remediation be needed later. In this case, future use of the site would come into play, especially if expansion of the manufacturing footprint is being considered and the "as is" section is used for that purpose.

The experiences at Newport, Delaware, and in Northern Ireland, stimulated another practice we incorporated into our corporate approach at other similar sites. That practice was to bring in ecologists and wildlife experts to survey these "as is" habitats to help us understand what ecological resources were present. In some cases, we also asked these individuals to determine if the "as is" section was suitable for restoration or for donation to a local or national conservation group.

Discussion

It is difficult to compress insights from a 34-year career into a manuscript and make it interesting enough to read. I may have failed, but hopefully there are some points that readers will find useful.

I saw many changes over time and was afforded the opportunity to work on some large and challenging issues, while also learning about how environmental toxicology was practiced in DuPont and other large companies inside and outside the United States.

I'd like to turn back to the three trends identified in the Introduction: (1) the beginning and subsequent widespread use of ecological risk assessment and risk management; (2) climate change and its influence on the practice of environmental toxicology and chemistry; and (3) remediation and restoration of contaminated land/habitats. I was fortunate to be involved with the development and subsequent use of the ecological risk assessment framework that spread throughout regulatory agencies and companies. I believe the guidelines and the many papers published on their application over the past 30 years is a laudable achievement. One of the hallmarks of the development and testing of the guidelines was the interactions between government, industry, and academia. Over time, these same sectors have been working to refine them and their use in a variety of situations. A few years ago, my colleague Dr. Will Gala (Chevron, retired, personal communication) remarked that the 1990s was "the golden age of eco-risk." I won't argue with that view, but I think the future of eco-risk is very bright. I think the practice will continue to evolve to deal more effectively with multiple stressors, probabilistic estimations, and climate change, among others.

The issue of climate change took hold in environmental toxicology only in recent times, thanks in part to the concern that the chemical stressors that most of us focus on could be even more problematic when coupled with the effects of a warming or drier climate (Wenning et al. 2010). Similar to efforts on ecological risk assessment, the interaction of climate change with chemical stressors has risen as an issue for environmental toxicologists and chemists within the SETAC community. Publications in the SETAC journals have highlighted the importance of the issue, and it has been the focus of two Pellston© Workshops – one in 2011 and another that was held in June 2022 (Stahl et al. 2013). The first workshop was designed to explore the interactions between chemical stressors and climate change, whereas the second workshop was designed to develop methods whereby climate change projections could be incorporated into ecological risk assessment. Publications from the first workshop appeared in *Environmental Toxicology and Chemistry*, while publications from the second workshop were published in the journal *Integrated Environmental Assessment and Management*, in Spring of 2023.

During my career, there have been substantial interactions between industry, government, and academia on environmental toxicology issues, manifested in conferences and workshops, many sponsored under the SETAC umbrella. The tripartite nature of SETAC made it a valuable, neutral platform where diverse technical views and needs could be discussed in an open, nonconfrontational manner. I hope this continues and provides the same opportunities for dialog to the new crop of environmental toxicologists in government, industry, and academia. I'm sure there will be new, as-yet unknown challenges for environmental toxicologists in the future, and this bodes well for continuing to support neutral platforms for discussing ways to address them.

Acknowledgments

I am indebted to my many coworkers in DuPont as well as those many professionals in other companies, government, academia, and nonprofits that I have worked with and gotten to know over my career. Each, in some way or the other, helped me on technical matters as well as provide training on how to work and communicate with people better.

Conflict of Interest Statement

The author did not receive funding from any public or private entity and declares no conflict of interest. All statements in the chapter are those of the author and not of any public or private entity mentioned herein.

Disclaimer

The views and opinions expressed herein are those of the author alone.

Data Availability

No specific data was used to prepare this manuscript.

References

Department for Environment, Food and Rural Affairs (DEFRA) (2004). *Model Procedures for the Management of Land Contamination*, Contaminated Land Report 11, 192. Bristol.

Department for Environment, Food and Rural Affairs (DEFRA) (2006). DEFRA circular 01/2006. Environmental Protection Act 1990, Part 2A. London, England. 190p.

Dunford, R.W., Ginn, T.C., and Desvousges, W.H. (2004). The use of habitat equivalency analysis in natural resource damage assessments. *Ecol. Econ.* 48: 49–70.

Kapustka, L.A., Bowers, K., Isanhart, J. et al. (2016). Coordinating ecological restoration options analysis and risk assessment to improve environmental outcomes. *Integr. Environ. Assess. Manag.* 12 (2): 253–263.

Kovalick, W.W. and Montgomery, R.H. (2017). Models and lessons for developing a contaminated site program: an international review. *Environ. Technol. Innovation* 7: 77–86.

Landis, W.G., Ayre, K.K., Johns, A.F. et al. (2017). The multiple stressor ecological risk assessment for the mercury-contaminated South River and upper Shenandoah River using the Bayesian network-relative risk model. *Integr. Environ. Assess. Manag.* 13 (1): 85–99.

Stahl, R.G. (2022). Science lessons from a mercury site. *Environ. Toxicol. Chem.* 41 (3): 497–498.

Stahl, R.G. Jr., Hooper, M.J., Balbus, J.M. et al. (2013). The influence of global climate change on the scientific foundations and applications of environmental toxicology and chemistry: introduction to a SETAC international workshop. *Environ. Toxicol. Chem.* 32 (1): 13–19.

Stahl, R.G. Jr., Gouguet, R., DeSantis, A. et al. (2008). Prospective environmental restoration/restoration up front: a concept for an incentive-based program to increase restoration planning and implementation in the United States. *Integr. Environ. Assess. Manag.* 4 (1): 6–14.

Stahl, R.G., Nicholson, M.J., Butler, P.B. et al. (2001). Managing ecological risks and restoration of a wetland habitat at a superfund site: a retrospective assessment of the proposed ecological risk management framework. In: *Risk Management: Ecological Risk-Based Decision Making* (ed. R.G. Stahl, R. Bachman, A.L. Barton, et al.), 149–155. Pensacola, FL: SETAC Press.

Taylor, R.A. (2007). *Policy and Practical Considerations in Implementing Restoration Banking*, 6. Washington, DC: National Oceanic and Atmospheric Administration.

U.S. Congress (1976). The resource conservation and recovery act. 42 United States code 6901. Washington, DC.

U.S. Congress (1980). The comprehensive environmental response, compensation and liability act (CERCLA) public law 95-510. Washington, DC.

U.S. Department of the Interior (USDOI) (2017). Code of federal regulations. Title 43 Subpart A,B,C,D,E,F. Part 11. Washington, DC.

U.S. Department of the Interior (USDOI) (2021). Best practices for considering advance restoration and restoration banking in the context of natural resource damage assessment and restoration, US Department of the Interior Restoration Program Authorities, Guidance, and Policy Library, US Department of the Interior, 2021. https://www.doi.gov/sites/doi.gov/files/best-practices-for-advanced-restoration-final-signed-june-11-2021.pdf.

U.S. Environmental Protection Agency (USEPA) (1992). *Framework for Ecological Risk Assessment*. EPA 630/R-92/001, 56. Washington, DC: Risk Assessment Forum.

U.S. Environmental Protection Agency (USEPA) (1993). USEPA record of decision. Newport Superfund Site, New Castle County, Delaware. EPA/ROD/R03-93/170. Philadelphia, PA.

U.S. Environmental Protection Agency (USEPA) (1995). *Ecological Risk: A Primer for Risk Managers*, 36. Washington, DC: USEPA Office of Prevention, Pesticides and Toxic Substances.

U.S. Securities and Exchange Commission (USSEC) (2020). Form 10-K. Annual report pursuant to Section 13 or 15(d) of the securities exchange act of 1934. General Instructions. Washington, DC, p. 12.

Wenning, R.J., Finger, S.E., Guilhermino, L. et al. (2010). Global climate change and environmental contaminants: a SETAC call for research. *Integr. Environ. Assess. Manag.* 6 (2): 197–198.

10

Big Data – Artificial Intelligence and Machine Learning – Examples of the Application of Approaches Derived from Artificial Intelligence/Machine Learning to Analyze Large Datasets for Ecological Risk Assessment

Wayne G. Landis

College of the Environment, Institute of Environmental Toxicology and Chemistry, Western Washington University, Bellingham, WA 98229, USA

Introduction

A key issue with conducting ecological risk assessments has been the difficulty in describing the effects of a chemical or other alterations to a dynamic system defined by nonlinear interactions, across a physical landscape and that changes according to natural selection (Landis et al. 2017, Chapter 3). By the publication of the description of risk assessment as used in the federal government, nicknamed the Red Book (NRC 1983), evidence was in place that ecological systems had the properties of complex systems, meaning that it was clear that they are not in equilibrium and the dynamics were nonlinear (in the vernacular "chaotic") and deeply interconnected. A property of nonlinear systems is that they are historical, that the past is not erased in the reaching of an equilibrium, but that the historical information continues to affect future dynamics.

The nonequilibrium nature of ecological structures and the dependence on the landscape is best summarized by Wu and Loucks (1995). In this keystone paper, the conventional models of the late 1980s to early 1990s were summarized (balance of nature, equilibrium static stability, nonequilibrium/instability, multiple equilibria/homeorhesis) and compared to hierarchical patch dynamics. The hierarchical patch dynamics was proposed as new and powerful paradigm to describe the new approach and is now referred to as the hierarchical patch dynamics paradigm (HPDP) (Landis et al. 2017). The implications of the HPDP in understanding the risk assessment process have been noteworthy. The assumptions that ecological systems can be described by equilibrium-based dynamics independent of history, and changes in the landscape are not relevant to predicting the probability of a specific outcome after either a single event or during a continuous alteration due to an episodic chemical spill or the current long-term climate change event.

In the 1980s, it was an assumption that toxicity testing for ecosystems could be done by means of an exposure response model using nondosed replicates (also called controls – but I cannot do that for reasons that will become clear) and several concentrations chosen to describe an exposure–response relationship. The design mimicked the standard model for toxicity testing using organisms at several specific concentrations in identical containers with three to five replicates per concentration. The data analysis was as typical in the 1980s, an analysis of variance hypothesis test followed by a multiple comparisons test. The goal was to find the concentration that was the highest that did not show a significant difference from the nondosed treatment and the lowest concentration that found a significant difference. Because these tests contained a variety of organisms in order to mimic the trophic levels and interactions of ecosystems, a variety of measurements were made, such as the numbers of each species and an index to represent diversity, among others. Because of the numbers of species used in these studies and the measurement of chemical concentrations and water quality parameters, the result was the generation of large datasets, especially for this era.

As will be demonstrated later in Section 1, the use of conventional statistics was not able to detect the persistence of the effects of contaminants even beyond degradation of the material. However, newer methods did.

In the late 1980s and through the 1990s, methods to perform ecological risk assessment of pesticides, effluents, metals, and other materials were developed (EPA 1992). A framework was developed that is still the basis for the risk assessment process for many regulatory bodies to the present day. At this time, the majority of toxicity results were reported as results from hypothesis testing as Lowest Observed Adverse Effect Concentrations (LOECs), No Observed Adverse Effect Concentrations (NOECs), or other summary statistics. In some cases, regression results were reported, but the experiments were based on the assumption that hypothesis testing were the primary data analysis approach. Risk was calculated as a quotient of an expected or measured environmental concentration over a number expressing an acceptable effect level such as an LOEC or NOEC. Moore and Caux (1997) documented using extensive datasets the limitations of hypothesis testing in detecting and describing toxicity. At the writing of the initial framework and guidance documents, robust tools developed for machine learning such as Bayesian networks (BNs) and structural equation modeling were not generally available. Estimation of risk due to multiple stressors in one environment was not supported in these earlier attempts to formulize risk assessment approaches.

In the mid-1990s, methods to calculate ecological risk at large scales and multiple stressors had been developed. Data analysis tools to identify patterns in datasets become more common. In the late 1990s and early 2000s, BNs were suggested as a means to calculate environmental risk. In the 2010 timeframe, the use of BNs had grown in risk assessment and other fields. The coupling of BNs to adaptive management was the next step. This chapter summarizes these events in three sections. The first is a review of the microcosm studies that spawned the use of machine learning–derived tools for data analysis. The second section summarizes two case studies, one centered on the fjord of Port Valdez, Alaska and the other on estuaries surrounding Brisbane, Australia. These cases illustrate the application of BNs and machine-learning approaches to calculate risk from multiple stressors to multiple endpoints. The last section discusses how these tools can be key to adaptive management.

Section 1: Microcosm/Mesocosm Summary, the Effects of Jet Fuels, and Community Conditioning

Several multispecies toxicity studies led to the use of a variety of approaches applicable for the analysis patterns within large datasets. For examples of this discussion, we will use two types of microcosms, the standardized aquatic microcosm (SAM) and the mixed flask culture (MFC) (Figure 10.1). Both underwent testing by multiple labs and were codified with published standard methods. Both methods were used by this author, and the examples presented are from this published research. It is this research and a collaboration with G. B. Matthews, R. A. Matthews, and our research labs that led to the adoption of several multivariate approaches and techniques that are part of the artificial intelligence sphere for the analysis of these types of experiments and ultimately the risk assessment of large-scale ecological systems.

Multispecies toxicity tests (microcosms) are the experimental design closest in generating very large datasets and are those that mimic the species diversity and dynamics of field systems. These microcosm/mesocosm studies involve multiple species at several trophic levels and at scales from 1 L or a few square meters to those resembling farm ponds. The Experimental Lakes region in Canada has been the site of many groundbreaking research programs and involves dividing the lakes into experimental areas with specific treatments. An outcome of these types of experiments are the large datasets.

For example, the standardized aquatic microcosm developed by Taub and colleagues (Taub and Read 1982; Taub 1989, 1989) can generate 10,686 datapoints from simply the counts of the different species, some basic environmental data, 24 experimental units, and 63 days of duration. This total does not count the data following the degradation of the water-soluble fraction (WSF) of the jet fuel added in three treatments (Landis et al. 1993a,b, 1994, 1995). Data analysis was originally based on the analysis of the dynamics of each of the measured parameters (counts of species, pH, chemical concentration, etc.). Intervals of nonsignificant difference (INSD) were calculated to indicate when and in what test treatments the

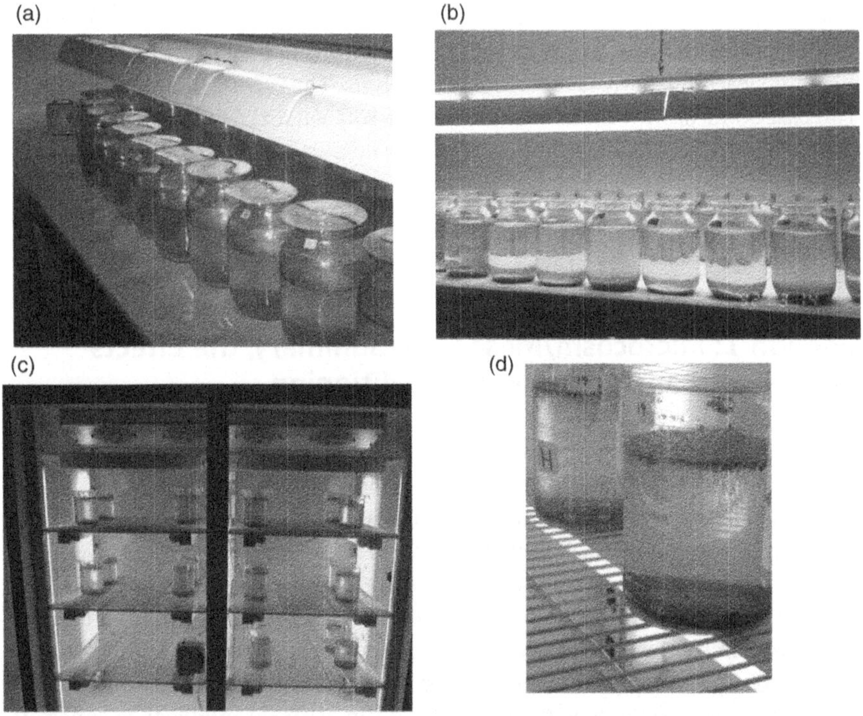

Figure 10.1 Standard aquatic microcosm (SAM) and mixed flask culture (MFC) multispecies toxicity test methods. (a) is a depiction of the initial stages of the experiment after the containers have been inoculated with the species listed in the protocol and randomized in the experimental room. (b) depicts a time in the experiment with the algae blooming due to the lack of Daphnia because of the toxicity. (c) depicts the arrangement of the individual MFC containers. *Source:* Landis et al. (2018) / with permission of Taylor and Francis. In (d), the cultures contain a variety of species that exist from the naturally derived inoculate.

values exceeded those in the nondosed treatment (control) replicates. This section describes how these types of experiments led to the adoption of multivariate techniques to find patterns in exposure–response interactions and the importance of these tools in formulating the community conditioning hypothesis.

The microcosm experiments followed the Taub (1989) protocols for the experimental design, species composition, measurement, and statistical analysis. The toxicant was the WSF of JP-4 and other jet fuels (Jet A, JP-8). Dosing was based on acute toxicity testing of the WSF. Treatment groups were 0%, 1%, 5%, and 12% WSF. After dosing with the jet fuel, they rapidly degraded and by week 3 (21 days) only 2-methlypropane could be detected. In several of the measured parameters, there were treatment-related and statistically significant results observed. However, using the analysis of variance–based INSD approach, there were no statistically different treatments. This pattern was observed in other jet fuel experiments using the Taub method.

The data analysis for these experiments included two multivariate clustering techniques, cosine distance and vector distance, and nonmetric clustering using the program Riffle. The question that these methods addressed was, Could we detect treatment differences in treatments based on community composition and other measured variables? In contrast to INSD analysis, treatment groups could be identified better than chance at the end of these experiments. In other words, the effects of the treatments could be detected even though the vast majority of the toxicant had been degraded. These treatment-related groupings were also found in experiments using Jet A, JP-4, and JP-8 (Landis et al. 1993). A common pattern was that early in the experiment treatment-related differences were clear, then a period midway in the 64-day time period the groupings were not statistically significant. Then later in the experiment, the treatment-related groups were again detectable (Figure 10.2). This pattern demonstrated that that information regarding treatment still existed within the microcosms even if it was not detectable for a period.

The nonmetric clustering method also identified the variables important in identifying the treatment groups. For example, on day 11 the variable *Selanastrum* and medium *Daphnia*, *Chlorella*, and *Ankistrodesmus* were those that best identified treatment groups. At day 60, and when the system was not as dominated by green algae and detritus driven, small *Daphnia*, *Ostracod*, and *Lyngbya* were key.

The inability of the univariate INSD to detect treatment-related differences later in the experiments that were detected by the multivariate methods was replicated using the other jet fuels in the Taub microcosm and in the MFC (1 L) systems (Sandberg and Landis 2001). These results led to the formulation of the community conditioning hypothesis (Matthews et al. 1995) and its confirmation using the results from microcosm toxicity tests (Matthews et al. 1994; Landis et al. 2000).

Visualization of the dynamics of the treatments groups is helpful in grasping the results of the experiments and their implications. In the microcosm experiments, this is particularly useful.

This visualization process involves projecting two variables over time in a three-dimensional graphical display (Figure 10.3). For any given sampling day, the mean for each treatment group is plotted. Around this mean is a circle with a radius proportional to the average distance between replicates within the group. The plots for each sampling day are then connected to give a worm-like projection showing the dynamics of the two plotted variables over time for each treatment (Landis et al. 1996). An example of the utility of this type of visualization in summarizing large datasets is presented next.

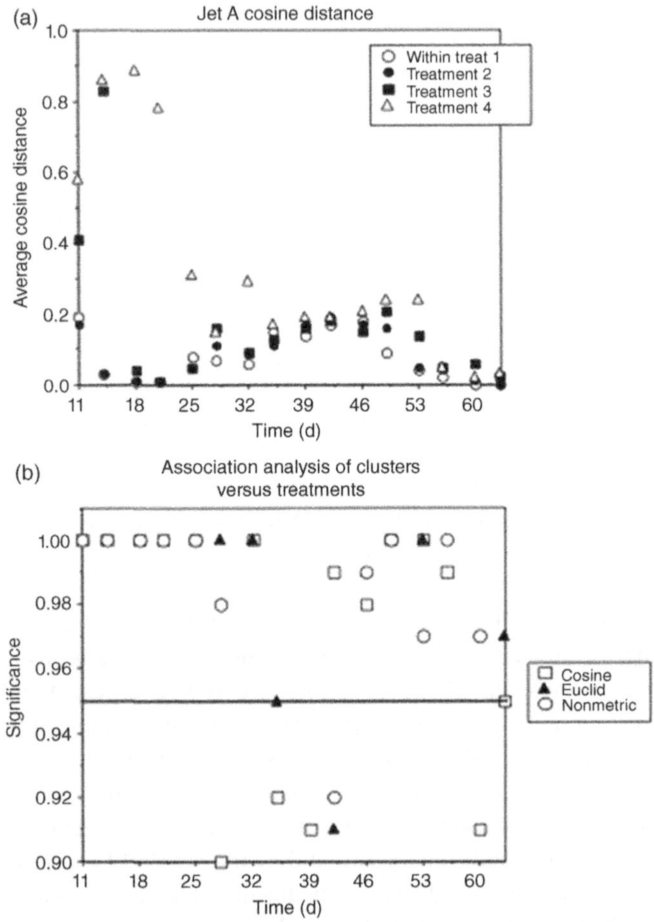

Figure 10.2 Multivariate analysis of the impact of Jet-A in the standardized aquatic microcosm (SAM) test system. (a) shows the Cosine distance from the control group to each of the treatments for each sampling day. Note that large differences are apparent early in the SAM. During the middle part of the 63-day experiment, the distances between the replicates of Treatment 1, the control group, is as large as the distances to the treatment groups. However, later in the experiment, the distances from the dosed microcosms to the control again increase. Significance levels of the three multivariate statistical tests for each sampling day are presented in (b).

Space-time worm projection of *A. cylindrica* and small amphipods are shown in Figure 10.3a. *A. cylindrica* shows treatment-related effects shortly after dosing with JP-8 with greater quantities of algae in the treated microcosms. This pattern persists until day 63 (marked by the torus). At this point, small amphipod populations begin to show distinguishing dynamics. By day 91, the amphipod populations are

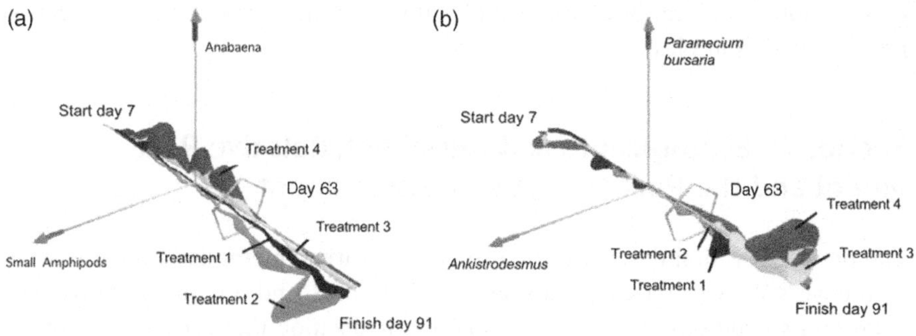

Figure 10.3 Trajectories of different treatments of JP-8 and heat in a Taub microcosm. The pentagon marks day 63 of a typical study. (a) is a space-time worm projection using *Anabena* and small amphipods as the variables. (b) depicts the trajectories for *Paramecium bursaria* and *Ankistrodesmus*. Note how the two trajectories provide information about each system at different times of the experiment. At certain periods of the experiments, the distance along these axes are small between treatment groups but can then change as the treatments age. Key: Reference are nondosed (Treatment 1), heat only (Treatment 2), JP-8 (Treatment 3), and JP-8/heat (Treatment 4).

distinct in each treatment (nondosed [blue], heat only [green], JP-8 only [yellow], and JP-8/heat [red]).

The space-time worm projection using *A. falcatus* and *P. bursaria* populations (Figure 10.3b) again shows two groups just after dosing with jet fuel based on *A. falcatus* populations. This is followed by a long convergence period with little variance within treatments through day 63. At this point, marked by the torus, the treatments become distinct from each other based on *P. bursaria* populations. The JP-8/heat-only treatment shows the greatest number of *P. bursaria* and the greatest variance within a treatment.

The community conditioning hypothesis states that ecological communities tend to preserve information about every event in their etiology (Landis et al. 2000). As that complex systems are historical, they are the result of the physical, chemical, and biological events. Therefore, no two ecological systems will be identical. As a consequence, the information regarding prior stress events can be kept in a variety of forms, from the genetics of the populations to types of detritus and community interactions. Recovery, as return to an initial condition, means that this information is lost as return to that condition eliminates that history. Therefore, recovery can be defined only by reference to a set of values as determined by culture.

The mid-1990s was a pivotal time in ecological risk assessment. It was then that questions about how to apply this decision making tool to large systems, with multiple stressors and multiple endpoints of concern to stakeholders, and how to apply the reality of nonequilibrium dynamics arose. The next section provides two examples.

Section 2: Ecological Risk Assessment/Relative Risk Model and the Port Valdez Risk Assessment

The previous section described the use of multivariate methods in the discovery of the persistence of treatment-related effects in 3 and 1 L microcosms. The community conditioning hypothesis and the realization that ecosystems are not equilibrium structures (Wu and Loucks 1995) has important ramifications for risk assessment. This section illustrates those ramifications to ecological risk assessment.

To illustrate those changes, this section involves a massive change in scale. Since the late 1990s, my research group and others have conducted risk assessments to calculate the risk of chemicals, oil spills, invasive species, stormwater runoff, and climate change. Instead of laboratory tests, the scales range from small watersheds to large segments of the United States and Australia. At the same time that it became apparent that multivariate approaches are needed to understand the impacts of chemicals and that large datasets were now the norm, the direction of the research lab became oriented toward the estimation of risk to large and diverse sites across the world. The first site involved was the Fjord of Port Valdez, and this initiated the formulation of the relative risk model (RRM). Examples of other projects are listed in Table 10.1.

Port Valdez is the terminus of the Alaska Pipeline transporting oil from the North Slope. Because of permitting process and the public interest, a large number of datasets and studies were available for the site. This included datasets for the occurrence of numerous species, water quality and other physical properties, benthic community structure, contaminants such as polycyclic aromatic hydrocarbons (PAHs), and so on. Large tankers transversed the fjord to be loaded with crude oil for transport to refineries. As described in Wiegers et al. (1998), there were numerous stakeholders and stakeholder groups with an interest in the site and its long-term management. The Regional Citizen's Advisory Committee (RCAC), which provides citizen oversight for the pipeline, funded the risk assessment program. The RCAC also acted as the connection between the risk assessment team and the numerous and diverse stakeholders of the Port Valdez study site.

The tsunami from Good Friday Earthquake of 1964 came ashore at Valdez, causing damage to the town and the port and altered the marine landscape. The port can be used year round and has a fishing fleet, a refinery, and a large tourist industry (Wiegers et al. 1998). Because of the history, several studies were available as datasets making possible the integration of a large dataset for its time. The RCAC

Table 10.1 Examples of ecological risk assessments using the relative risk mode (RRM), probabilistic relative risk model (PRRM) using Monte Carlo, and the Bayesian network–relative risk model (BN-RRM) as noted in the text.

Site	Stressors	Endpoints	Approach	Citations
Fjord of Valdez, Alaska	Hydrocarbons, metals, organic matter and nutrients, hatchery fish	Water and sediment quality, finfish and shellfish populations, wildlife populations such as fishes, birds, and mammals	Relative risk model	Wiegers et al. (1998)
Cherry Point, Ferndale-Blaine, Washington	Behavioral and physical disturbances, contaminants, exploitation, sedimentation patterns, shading of near shore vegetation, removal of terrestrial habitat	Coho salmon, juvenile Dungeness crab, juvenile English sole, great blue heron, littleneck clam, surf smelt	Probabilistic relative risk model	Hart Hayes and Landis (2004)
INLAS, Oregon	Grazing, insects, wildfire, forest management	Timber, grazing, recreation, HRV wildfire, insects, invasives, salmon habitat	Probabilistic relative risk model	Anderson and Landis (2012)
INLAS, Oregon	Grazing, insects, wildfire, forest management	Timber, grazing, recreation, HRV wildfire, insects, invasives, salmon habitat	Bayesian network–relative risk model	Ayre and Landis (2012)
South River, Waynesboro, Virginia	Mercury, PAHs, organochlorine pesticides. Water quality criteria for temperature, total suspended solids, avian nest predation, dissolved oxygen, total phosphorus, habitat quality, and discharge regime	Smallmouth bass, white sucker, belted kingfisher, Carolina wren, water quality standards compliance, fishing river use, swimming water use, boating water use	Bayesian network–relative risk model	Landis et al. (2017)
Estuaries in the Brisbane, NSW, Australia	Dissolved oxygen, chlorophyll A, benthic community structure as measured by eDNA	Water temperature, turbidity, total nitrogen, total phosphorus, surface water salinity	Bayesian network–relative risk model	Graham et al. (2019)
Yakima River (Columbia River basin)	Pesticides, water quality parameters	Chinook Salmon	Bayesian network–relative risk model	Landis et al. (2020), Mitchell et al. (2020), Landis et al. (2024)

HRV, Historic Range of Variability; INLAS, Interior Northwest Landscape Analysis System; PAH, polycyclic aromatic hydrocarbon.

also facilitated local public, private, and governmental participation and cooperation in obtaining additional information.

As first presented in Landis and Weigers (1997) and then fully realized in the ecological risk assessment for Port Valdez (Wiegers et al. 1998), the RRM is a structure that was amenable to the use of a variety of techniques to inform the ranking process and the links describing cause–effect pathways.

The management question of the RCAC was the treatment of 21 million gallons of treated ballast water from the tankers that are preparing to transport crude oil from the pipeline. Ballast water treatment plant (BWTP) discharges the treated water under a National Pollution Discharge Elimination System (NPDES) for levels of hydrocarbons.

However, there were also a diverse set of sources of contaminants from other areas of the fjord that may compound to cause declines in valued populations and affect water quality when combined with the BWTP effluent. The ecological risk assessment frameworks available in the mid-1990s was not designed to incorporate several different stressors coming from a wide variety of sources and to integrate their effects to various endpoints. The RRM was developed to incorporate a wide variety of stressors coming from a wide variety of sources so that risk to multiple endpoints could be calculated. The RRM also included methods for describing uncertainty and the study of what-if scenarios to investigate suggested mitigation activities.

The basic structure of the risk assessment also incorporates the fundamental cause–effect relationship required for a risk assessment (Figure 10.4). Ranks were used to describe the sizes of sources, the stressors, the amount of habitat for a specific species, the exposure–response (effects) for each endpoint, and finally the risk for each. The use of ranks allowed the multiple sources, stressors, and so on to be integrated regardless of units. The use of discreet ranks made it straightforward for the eventual transition to BNs for the process in our follow-on research.

As summarized in Wiegers and Landis (2004), the RRM was successful in estimating risk from multiple sources of multiple types from across the fjord. Uncertainty was quantified into five types:

1) Missing information
2) Ambiguities in available information
3) Errors in the conceptual model

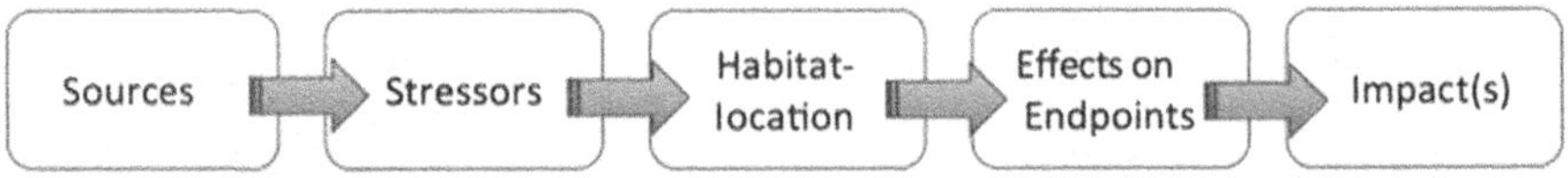

Figure 10.4 The basic format of the relative risk model (RRM). This format is still the basis of the current Bayesian network-based RRMs.

4) Errors in the estimate of relative risk
5) Variability in the environment.

The sensitivity of the findings was also characterized given these types of uncertainties and methods were proposed to confirm the risk predictions.

There are three outputs that still remain the standard for the RRM approach:

1) Maps of the risk regions with the associated sources, land uses, habitats, and the spatial distribution of the assessment endpoints.
2) A regional comparison of the relative risks, their causes, the patterns of impacts to the assessment endpoints, and the associated uncertainty. These regional comparisons and estimates of the contribution of each source and stressor creates a spatially explicit risk hypothesis. Further field and laboratory research or analysis of the existing database can be used to test these hypotheses.
3) A model of source-habitat-impact that can be used to ask what-if questions about different scenarios that examine potential options in environmental management. It was straightforward to change to the model to reflect a reduction in a specific source of contamination.

Risk scores varied geographically within the study area. Systematic application of the conceptual model to the habitats and risk sources in each of the subareas led to a ranking of relative risk within the port environment. The risk scores are unitless numbers that judge the relative severity of environmental risk based on an informed decision-making process. The higher the relative risk score, the higher the risk.

Risk can vary dramatically among sites in the same system. The scores ranged from 40 (Sawmill to Seven-Mile Creeks) to 448 (Duck Flats and Old Valdez). We considered subareas with scores less than 150 to have low relative risk. Subareas in this group included Shoup Bay, Sawmill to Seven-Mile Creeks, Anderson Bay, and the Western Port. Subareas with scores between 150 and 300 were considered to have moderate relative risk. These included Mineral and Gold Creeks, City of Valdez, Robe and Lowe Rivers, Dayville Flats and Solomon Gulch, and the Valdez Marine Terminal. Only one subarea, Duck Flats and Old Valdez, had a high risk score greater than 300. Because of the uncertainty associated with the data collection, toxicity data, and the relative ranking process, comparisons of relative risk with these categories is not meaningful.

The analysis suggested that the pelagic environment and western shoreline, areas affected by less development, are at low relative risk. Most of the eastern shoreline is at moderate relative risk. This includes subareas from the City of Valdez to the Valdez Marine Terminal where a variety of development has occurred. The one subarea of high relative risk, Duck Flats and Old Valdez, is located in the developed eastern area. The greater risk predicted here by the model is related to the diversity and quality of habitats in this area. Note that "high relative risk" may or may not imply high risk in an absolute sense. Instead, this suggests that a greater degree of environmental stress is possible and that there is a higher probability that significant ecological impacts will occur than in other areas of the port.

The contribution of the eight-stressor sources to relative risk in the entire Port Valdez region was also calculated. Applying the same criteria defined earlier (low relative risk, <150; moderate relative risk, 150–300; and high relative risk, >300), treated discharges, fish and seafood wastes, and the presence of hatchery fish rank as low relative risk; vessel traffic and construction and development activities as moderate relative risk; and contaminated runoff, accidental spills, and shoreline activity as high relative risk. This distribution of relative risk between sources is reasonable when characteristics of the stressors associated with these sources is considered. Runoff, spills, and shoreline activity behave similarly to nonpoint discharges, and the effects are likely to be widely distributed throughout the port. Treated discharges and fish and seafood wastes behave more like point sources and may be discharged into fewer subareas and possibly less-sensitive habitats.

The contribution of the eight habitat categories to relative risk in Port Valdez were also calculated. Using the same criteria defined earlier, saltmarsh and deep benthic habitats rank as low relative risk; spits and low-profile beaches, the rocky shoreline, and open water habitats as moderate relative risk; and mudflats, shallow subtidal, and stream mouth habitats as high relative risk. Relative risk to habitats in Port Valdez as a whole is strongly influenced by the abundance of habitats across subareas. For instance, saltmarsh occurs in only one subarea, Duck Flats and Old Valdez. Although saltmarsh receives the highest possible ranking *in that subarea*, that alone still leads to a low relative risk to Port Valdez *as a whole*. The reverse situation occurs for open water habitat. The risk to open water in any individual subarea is never more than half the maximum possible, but open water occurs in every subarea. The result is that open water habitats have high relative risk for Port Valdez as a whole. Landis and Wiegers (1997) have been cited over 300 times in diverse ecological risk assessment studies at sites across the world.

Improvements to the RRM method now include the incorporation of Monte Carlo simulation to capture uncertainties to calculate risk distributions and to examine the sensitivity of these distributions to the inputs (Hayes et al. 2004). The use of ranks to produce probabilistic estimates applied to multiple stressors and endpoints across landscapes had been demonstrated. The biggest change was the transition to the use of BNs (also called Bayesian belief networks) to estimate risk, calculate sensitivity, and quantify uncertainty.

Section 3: Transition to BNs

BNs are directed acyclic graphs comprising parent nodes that depicts the inputs of two variables, a child node that shows the results, and a conditional probability table (CPT) that describes the interactions between the parent nodes. Each parent node is a probability distribution of discrete states. The child node is a probability distribution of outcomes, again described as discrete states. Kaikkonen et al. (2021) extensively reviewed the literature describing the use of BNs in environmental assessment. The transition to BNs to calculate relative risk began in the mid-2000s.

In the late 2000s, it became apparent that BNs could be applied to risk assessment and especially the RRM. Anderson and Landis (2012) presented a risk assessment for the Upper Grande Ronde Watershed using data and management objectives that were part of the Interior Northwest Landscape Analysis System (INLAS) project as managed by the US Forest Service. Decades of measurements had been integrated into detailed geographic information system (GIS) files, and clear management goals and management strategies existed. Unlike most risk assessments, this study did not focus on chemical contaminants but on factors that affected the forest. Instead, stressors included forest management, grazing, insects, and wildfire. The endpoints include the Historic Range of Variability (HRV) as the management goal for the area – in this study being pre-European settlement. HRV for the fire regime, invasive plants, and insects were selected. Forest resources were the key endpoints including recreation, hunting/fishing, grazing, and timber. Uncertainty and sensitivity analyses were also conducted. The results of the risk assessment were compared with the Prioritization of Washed Restoration Report (POWR) conducted by the Wallowa-Whitman National Forest. There were several similarities in the reports although conducted via different means. In comparison with the POWR report, the RRM approach could also identify the key variables in modeling the output (sensitivity) and cataloged the variables contributing to the uncertainty.

To compare the conventional RRM results to those using a BN, Ayre and Landis (2012) used the conceptual model and datasets of Anderson and Landis (2012) to build a BN version to calculate risk, estimate uncertainty, and calculate sensitivity. Although the methods are different in the approach to quantify the risk, they are similar in the risk estimates. Both approaches identified that the stressors with the highest probability to affect the endpoints were wildfire and forest management practices. The models predicted that the HRV fire endpoint would be at a substantial risk, but the HRV salmon habitat was the endpoint at highest risk according to the Bayesian-based model.

A particular advantage of the BN-based model was the ease of running a variety of different scenarios to examine the effects of various management plans to the endpoints. A single change in a management approach would affect several endpoints. Subtle changes in management approaches could greatly affect the distributions of several of the endpoints. The distributions of the endpoints were rarely normally distributed. The flexibility of the BN model led to its being the preferred approach for further calculations of relative risk approach.

The BN–relative risk model (BN-RRM) has now been applied to multiple kinds of stressors at a wide range of locations (Landis 2021). The transformation of the original RRM to the structure of BNs has allowed a better description of uncertainty, takes advantage of the computational flexibility of the BN network and the various software packages that incorporate it, and takes advantage of the extensive environmental assessment research using these methods. Uncertainty is easier to describe, sensitivity analysis is easy to calculate, and tools such as machine learning can be used to estimate exposure–response interactions (Kaikkonen et al. 2021; Landis 2021).

A specific drawback to the application of the BN-RRM is the requirement to learn to apply BNs to risk assessment. Since 2012, Netica, Bayeslab, BayesFusion, Hugin, and others have become available. The models in this chapter were developed using Netica because of the widespread use of the software at the time (early 2010s) and the fact that a free download was available to examine the models published at that time.

Case study 3 is an example of the application of BN–RRM and the versatility of the approach is the case study describing the risk to estuary water quality and to changes in ecological community structure as measured by eDNA (Graham et al. 2019). This study had access to thousands of data.

Section 4: Case Summary: Predicting Risk to Estuary Water Quality and Patterns of Benthic Environmental DNA in Queensland, Australia

This case summarizes the use of the BN-RRM to estimate the risk to water quality and DNA-derived benthic community structure for three estuaries in South East Queensland (SEQ), Australia. The sites include the Noosa, Pine, and Logan Estuaries, which have a range of anthropogenic disturbance: the Noosa with the least and the Logan with the most. CSIRO, Australia's national science agency, has collected environmental DNA (eDNA) from surface sediments in these estuaries and has determined that the patterns of eukaryotic composition reflect the environmental gradients in the estuaries (Chariton et al. 2014).

Environmental DNA analysis is a relatively new technique for biodiversity measurement that involves extracting DNA from environmental samples, sequencing a region of the 18s rDNA gene in eukaryotes, and matching the sequences to organisms. Putting eDNA into context with regional management objectives is the motivation for this project. The application of eDNA to risk assessment is largely unexplored; to our knowledge, this is the first attempt to synthesize DNA-derived measurements of biological composition into an ecological risk assessment framework. Risk to achieving water quality objectives was calculated using 15 years of site-specific monitoring data to quantify relationships between salinity gradients, rainfall, temperature, and land use to water quality and benthic community endpoints.

The conditional probability tables for each estuary BN-RRM were parameterized using case files and the expectation-maximization (EM-learning) algorithm, a function available in Netica® (Norsys 2014) that deals well with missing data. A case file is a compilation of datasets that go together to provide information about the variables (nodes) in the model, for example, water quality data combined with weather data collected on the same day. The EM-learning algorithm iteratively calculates the maximum likelihood estimates for the node states in the model given the case file data and the model structure.

An estuary-specific case file was used to parameterize the CPTs. The three case files consisted of Southeast Queensland Ecosystem Health Monitoring Program (SEQ EHMP) monitoring data that had been collected monthly from 1999 to 2014 at regular locations within the estuary. Each record (or case) in the case file was matched to the corresponding previous 30-day rainfall total and percentage intensive land use cases (n = 5,032 for Noosa, n = 6,204 for Logan, and n = 3,621 for Pine). These case files were used to parameterize the CPTs between the water quality endpoint nodes.

The benthic endpoint nodes of the model were estimated by using a single separate benthic data case file for all estuaries (n = 287). A single file was used because the eDNA dataset for individual estuaries was relatively small. Using benthic samples from different estuaries to parameterize CPTs is consistent with Lucena-Moya et al. (2015), who also used BNs and case learning to predict ecological assemblages.

Two other types of information were used to construct and parameterize the models. There is an extensive estuary science literature specific to this region. Second, there was close interaction with the scientists and environmental managers for these regions that provided information on context and the management goals for each site.

The BN for the Logan estuary is presented in Figure 10.5. The risk region is the initial node reading the model from left to right. This node specifies the datasets to be used to calculate the final results. The tan nodes are the variables that determine the water quality parameters, the green nodes. The blue nodes (last column top) are the water quality nodes with the yellow boxes outlining the management objective and the probability of meeting it. The orange nodes are the predictions

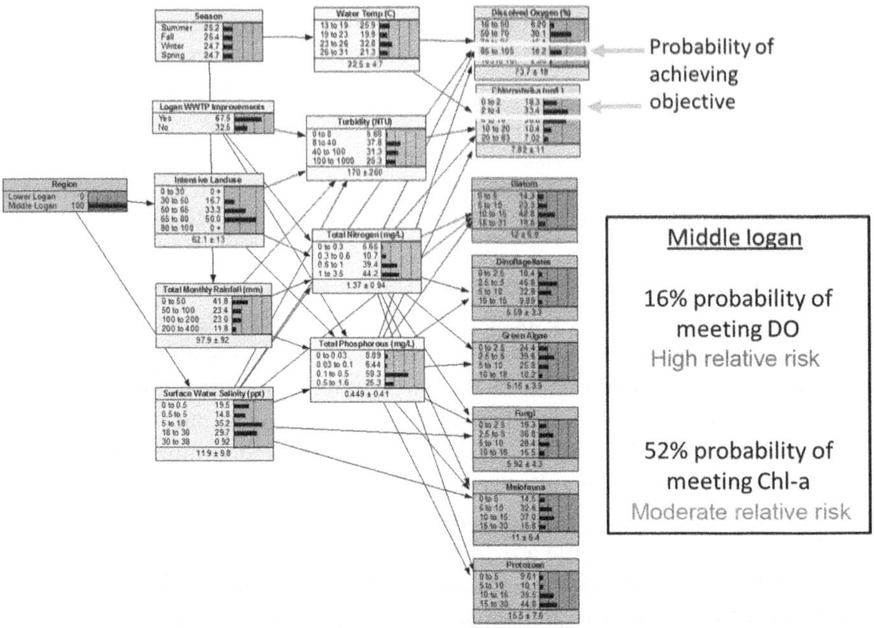

Figure 10.5 Bayesian network (BN) model example for the Logan estuary. A model was built for each of the study areas. This is the BN for the Lower and Middle part of the Logan with the Middle Logan selected for the risk calculation. *Source:* Graham et al. (2019) / with permission of Oxford University Press.

of occurrence of the frequency of the types of organisms as determined by eDNA. The risks to each of the endpoints for each of the risk regions in each estuary are detailed in Table 10.2. The probability of meeting the management goals for each site are listed. The Noosa estuary has the highest probability of meeting the management goals for each of its three risk regions meaning that it is at lowest risk. The risk regions in the Pine and the Logan had lower probabilities of meeting the management goals and therefore were at a higher risk compared with the Noosa. The regional specificity of the risk estimation was shown in both the Pine and the Logan, with clear differences between endpoints and locations.

Predictions from the model were compared with the results from sampling. The model is more accurate at predicting dissolved oxygen than it is the Chlorophyll-a water quality endpoint, and it predicts the diversity of photosynthesizing benthos more accurately than heterotrophs.

Results of BN-RRM modeling indicate that the water quality and benthic assemblages of the Noosa are relatively homogeneous across all subrisk regions and that the Noosa has a high probability (73–92% probability) of achieving water quality objectives, which indicates low relative risk.

Table 10.2 Probability of meeting management objectives to two endpoints by each of the risk regions.

	Dissolved oxygen	Chlorophyll-a	
Risk regions	**% Probability to achieve objective**		**Relative risk**
Noosa lower	74%	73%	Medium
Noosa middle	81%	77%	Low
Noosa upper	75%	91%	Low
Pine lower	55%	26%	Medium/high
Pine middle	24%	43%	High
Logan lower	69%	60%	Medium
Logan middle	16%	52%	High/medium

One characteristic of the risk assessments, especially those using the BNs, is that these models form the framework for designing studies and data analysis programs for an interactive adaptive management/decision making process. Wyant et al. (1995) proposed a framework for decision making for restoration activities. Van den Brink et al. (2016) proposed incorporating into this framework a variety of new data analysis tools and methods for calculating risk for an adaptive management process. Landis et al. (2017) took this proposal further using the South River risk assessment as a scenario.

Section 5: Adaptive Management and Next Steps

Wyant and Meganck (1995) proposed a general decision-making framework for quantifying the decision-making process for the restoration. The data analysis tools and the use of BNs applied to this general outline can operationalize the process. It should be possible to construct a similar framework using BNs as the computational background. Nyberg et al. (2001) discussed the use of BNs in adaptive management. In a classic study, Carriger and Barron (2011) explored the use of influence diagrams, that is, BNs that explicitly represent decisions related to an issue and the influence of those decisions on desired or undesired outcomes. Landis et al. (2017a) outlined how BNs could be used to conduct an interactive process for planning the study designs for laboratory and field research to acquire the appropriate data for building a BN suitable for incorporating existing knowledge, calculating risk, identifying uncertainties, testing the predictions, and then modeling remediation activities. As more data are acquired, the process can repeat, mitigation incorporated, results measured, and the learning from prior results built into a revised BN model. The South River study (Landis et al. 2017b) was used to illustrate the process. In other words, the adaptive management process uses a process similar to that in machine learning to learn as an organization how to effectively manage restoration. Cains and Henshel (2019) used Charleston Harbour and as an example for conducting Resilience Assessments to guide management activities. Cains has further described adaptive management as a convergence of multiple sciences including climate change (Cains et al. 2024).

Cautions/Caveats

The tools described here will allow a more complete understanding of how to manage the environment in a fashion the provides several advantages compared with current algorithms. However, there are issues.

The size of the datasets make it more challenging for peer review and outside verification of the analysis to occur. My research team has downloaded information from numerous governmental and other datasets from a variety of long-term monitoring programs and raw data from toxicity testing. Each are challenging to check to ensure that the transmission has been accurate, that the supplemental information is sufficient, and that the statistical and machine learning code has been verified against standard datasets. My laboratory routinely asks investigators for the raw data from a published experiment for our calculation of exposure–response curves. Summary statistics such as LOECs or EC50s without the datasets inherently underrepresent the uncertainty in the measurements and to not specify the number for concentrations tested. Likewise *P* values are published but rarely the statistical power of the experimental design. Fortunately, our collaborators and colleagues have been generous, even copying pages from laboratory notebooks, in answering our inquiries. For some information, the original data no longer exists, leaving only the summary.

We and others also make the models themselves available. Research from my lab routinely use Netica as the modeling software for our BNs. The free version allows anyone to read (but not modify) a .neta file. We now routinely add the model to the supplemental information for the study. Our models also come with notes/comments of assumptions and other aspects of the process. Many others do the same.

Another challenge will be in the use of large language models with proprietary code and learning datasets. It is not yet clear how these programs will be tested for reliability for the making of environmental decisions. Who will peer review the models, other AI models? It is not clear at this writing.

Having the predictions large language models tested will be crucial in the application of them in an adaptive management mode. How will stakeholders and decision makers ensure that the model outputs are reliable given their proprietary nature and commercial goals? Do the models represent a policy bias, unintentional or otherwise, by the developers? How would and how much trust could be applied to the outputs?

References

Anderson, S.A. and Landis, W.G. (2012). A pilot application of regional scale risk assessment to the forestry management of the Upper Grande Ronde watershed, Oregon. *Hum. Ecol. Risk. Assess.* 18: 705–732.

Ayre, K.K. and Landis, W.G. (2012). A Bayesian approach to landscape ecological risk assessment applied to the Upper Grande Ronde watershed, Oregon. *Hum. Ecol. Risk. Assess.* 18 (5): 946–970.

Cains, M.G. and Henshel, D. (2019). Community as an equal partner for region-based climate change vulnerability, risk, and resilience assessments. *Curr. Opin. Environ. Sustain.* 39: 24–30.

Cains, M.G., Desrousseaux, A.O., Boxall, A.B. et al. (2024). Environmental management cycles for chemicals and climate change, EMC4: a new conceptual framework contextualizing climate and chemical risk assessment and management. *Integr. Environ. Assess. Manag.* 20 (2): 433–453.

Carriger, J.F. and Barron, M.G. (2011). Minimizing risks from spilled oil to ecosystem services using influence diagrams: the Deepwater Horizon spill response. *Environ. Sci. Technol.* 45: 7631–7639.

Chariton, A.A., Ho, K.T., Proestou, D. et al. (2014). A molecular-based approach for examining responses of eukaryotes in microcosms to contaminant-spiked estuarine sediments. *Environ Toxicol Chem* 33: 359–369.

Graham, S.E., Chariton, A.A., and Landis, W.G. (2019). Using Bayesian networks to predict risk to estuary water quality and patterns of benthic environmental DNA in Queensland. *Integr. Environ. Assess. Manag.* 15: 93–111. https://doi.org/10.1002/ieam.4091.

Hart Hayes, E. and Landis, W.G. (2004). Regional ecological risk assessment of a nearshore marine environment: Cherry Point, WA. *Hum. Ecol. Risk. Assess.* 10: 299–325.

Kaikkonen, L., Parviainen, T., Rahikainen, M. et al. (2021). Bayesian networks in environmental risk assessment: a review. *Integr. Environ. Assess. Manag.* 17 (1): 62–78.

Landis, W.G. (2021). The origin, development, application, lessons learned, and future regarding the Bayesian network relative risk model for ecological risk assessment. *Integr. Environ. Assess. Manag.* 17: 79–94.

Landis, W.G. and Wiegers, J.A. (1997). Design considerations and a suggested approach for regional and comparative ecological risk assessment. *Hum Ecol Risk Assess* 3: 287–297.

Landis, W.G., Matthews, R.A., Markiewicz, A.J., and Matthews, G.B. (1993a). Multivariate analysis of the impacts of the turbine fuel JP-4 in a microcosm toxicity test with implications for the evaluation of ecosystem dynamics and risk assessment. *Ecotoxicology* 2: 271–300.

Landis, W.G., Matthews, R.A., Markiewicz, A.J. et al. (1993b). Multivariate analyses of the impacts of the turbine fuel jet A using a microcosm toxicity test. *J. Environ. Sci.* 2: 113–130.

Landis, W.G., Matthews, G.B., Matthews, R.A., and Sergeant, A. (1994). Application of multivariate techniques to endpoint determination, selection and evaluation in ecological risk assessment. *Environ. Toxicol. Chem.* 12: 1917–1927.

Landis, W.G., Matthews, R.A., Markiewicz, A.J., and Matthews, G.B. (1995). Non-linear oscillations detected by multivariate analysis in microcosm toxicity tests: implications for risk assessment and biomonitoring. In: *Environmental Toxicology and Risk Assessment: Third Volume*. ASTM International. Edited by J.S. Hughes, G.R. Biddinger, E. Mones.

Landis, W.G., Matthews, R.A., and Matthews, G.B. (1996). The layered and historical nature of ecological systems and the risk assessment of pesticides. *Environ. Toxicol. Chem.* 15 (4): 432–440.

Landis, W.G., Markiewicz, A.J., Matthews, R.A., and Matthews, G.B. (2000). Confirmation of the community conditioning hypothesis: persistence of effects in model ecological structures dosed with the jet fuel JP-8. *Environ Toxicol Chem* 19: 327–336.

Landis, W.G., Markiewicz, A.J., Ayre, K.K. et al. (2017). A general risk-based adaptive management scheme incorporating the Bayesian network relative risk model with the south river, Virginia, as case study. *Integr. Environ. Assess. Manag.* 13: 115–126.

Landis, W.G., Ayre, K.K., Johns, A.F. et al. (2017a). The multiple stressor ecological risk assessment for the mercury contaminated south river and upper Shenandoah river using the Bayesian network-relative risk model. *Integr. Environ. Assess. Manag.* 13: 85–99.

Landis, W.G., Sofield, R.M., and Yu, M.-H. (2018). *Introduction to Environmental Toxicology: Molecular Substructures to Ecological Landscapes*, Fifthe. Boca Raton, FL: Taylor Francis/CRC Press.

Landis, W.G., Chu, V.R., Graham, S.E. et al. (2020). Integration of chlorpyrifos acetylcholinesterase inhibition, water temperature, and dissolved oxygen concentration into a regional scale multiple stressor risk assessment estimating risk to Chinook salmon. *Integr Environ Assess Manag* 16 (1): 28–42.

Landis, W.G., Mitchell, C.J., Hader, J.D. et al. (2024). Incorporation of climate change into a multiple stressor risk assessment for the Chinook salmon (*Oncorhynchus tshawytscha*) population in the Yakima River, Washington, USA. *Integr. Environ. Assess. Manag.* 20 (2): 419–432.

Lucena-Moya, P., Brawata, R., Kath, J. et al. (2015). Discretization of continuous predictor variables in Bayesian networks: an ecological threshold approach. *Environ. Model. Softw.* 66: 36–45.

Matthews, G., Matthews, R., and Landis, W. (1994). Nonmetric clustering: new approaches for ecological data. In: *Proceedings of the Tenth Conference on Artificial Intelligence for Applications*, 482–483. IEEE.

Matthews, G., Mathews, R., and Landis, W. (1995). Nonmetric conceptual clustering in ecology and ecotoxicology. *AI Applications* 9 (1): 41–48.

Mitchell, C.J., Lawrence, E., Chu, V.R. et al. (2020). Integrating metapopulation dynamics into a Bayesian network relative risk model: Assessing risk of pesticides to Chinook salmon (Oncorhynchus tshawytscha) in an ecological context. *Integr Environ Assess Manag* 17 (1): 95–109.

Moore, D.R.J. and Caux, P.-Y. (1997). Estimating low toxic effects. *Environ. Toxicol. Chem.* 16: 764–801.

National Research Council (1983). *Risk assessment in the federal government: Managing the process*. Washington (DC): National Academies Press. https://doi.org/10.17226/366.

Norsys (2014). *NeticaTM*. Vancouver, BC: Norsys Software Corporation.

Nyberg, J.B., Marcot, S.R., Sandberg, R., and Landis, W.G. (2001). Persistence and the effects of jet-A in a sediment microcosm. *Environ. Toxicol. Chem.* 9: 1942–1950.

Sandberg, R. and Landis, W.G. (2001). Persistence and the effects of jet-A in a sediment microcosm. *Environ Toxicol Chem* 9: 1942–1950.

Taub, F.B. (1989). Standardized aquatic microcosms. *Environ. Sci. Technol.* 23: 1064–1066.

Taub, F.B. and Read, P.L. (1982). *Standardized Aquatic Microcosm Protocol*, vol. 2. 223-80-2352. Final Report. Washington, DC: Food and Drug Administration.

Taub, F.B., Kindig, A.C., Conquest, L.L., and Meador, J.P. (1988). Results of the interlaboratory testing of the standardized aquatic microcosm protocol. In: *Aquatic Toxicology and Hazard Assessment; 11th Volume, ASTM* (ed. G. Suter and M. Lewis), 368–394. Philadelphia, PA: American Society for Testing and Materials.

US Environmental Protection Agency (1992). *Framework for ecological risk assessment*. Washington (DC): USEPA Risk Assessment Forum EPA/630/R-92/001.

Van den Brink, P., Choung, C.B., Landis, W.G. et al. (2016). New approaches to the ecological risk assessment of multiple stressors. *Mar. Freshw. Res.* 67: 429–439.

Wiegers, J.K. and Landis, W.G. (2005). Chapter 4 Application of the Relative Risk Model to the Fjord of Port Valdez, Alaska. In: *WRegional Scale Ecological Risk Assessment Using the Relative Risk Model* (ed. G. Landis), 53–90. Boca Raton: CRC Press.

Wiegers, J.K., Feder, H.M., Mortensen, L.S. et al. (1998). A regional multiple stressor rank-based ecological risk assessment for the fjord of Port Valdez, AK. *Hum. Ecol. Risk. Assess.* 4: 1125–1173.

Wu, J. and Loucks, O.L. (1995). From balance of nature to hierarchical patch dynamics: A paradigm shift in ecology. *Q Rev Biol* 70: 439–466.

Wyant, G., Meganck, R.A., and Ham, S.H. (1995). A planning and decision-making framework for ecological restoration. *Environ. Manag.* 19: 789–796.

11

New Tools and Approaches in Ecological Risk Assessment

P. Bruce Duncan[1] *and Jeanmarie M. Zodrow*[2]

[1] *Bruce Duncan Environmental Consulting, Boulder, CO, USA*
[2] *Geosyntec Consultants, Risk Assessment and Applied Toxicology, Greenwood Village, CO, USA*

Introduction

In January 2019, one of us prepared an unpublished summary of emerging tools that several US Agencies had spent significant effort to develop or support. These tools showed great promise in helping ecological risk assessors in the EPA regions and national programs address issues of concern across the Ecological Risk Assessment paradigm (e.g., characterizing toxicity of unknown compounds, quantifying and communicating ecosystem importance, extrapolating effects across species, assessing cumulative effects, and using in situ approaches). Some of the tools selected were in use and were being enhanced, others had undergone successful demonstrations, and still others had passed a proof of concept and appeared ready for use in case studies. For this chapter, we had hoped to update and report on the kinds of decisions that had been made that relied on these tools. However, it became clear that some of these tools had yet to be incorporated into mainstream decision-making processes.

Despite the Goal to Advance Ecological Risk Assessment, Progress Can Be Very Slow

A goal of many scientific associations such as the Society of Environmental Toxicology and Chemistry (SETAC) is to advance environmental science and science-informed decision-making (SETAC mission statement). Scientists continually identify new and promising tools yet also note the lack of advancement (See Kapustka 2026). For example, Kendall and Akerman (1992) stated that understanding tissue uptake, distribution, and elimination of pesticides in wildlife would be fundamentally important to the evolution of wildlife toxicology as a science. Some 30 years later, Rattner et al. (2023) note that despite advances in toxicity testing and new approach methodologies (NAMs) for hazard assessment, "the ecological risk assessment framework for terrestrial wildlife has remained unchanged for decades." In some areas, inclusion of new tools can be rapid (e.g., more sensitive analytical methods). Given that some tools seem to take a long time before being used in decision-making, yet others are quickly successful, we compiled best practices for scientists developing new tools and approaches for ecological risk assessment.

Challenges in getting decision-makers to use the new tools have consistently been reported for a wide range of new approaches. One aspect is lack of integration of case studies with management decisions and reporting back. For example, in their review of eleven case studies of cumulative effects assessments (CEAs), Stelzenmüller et al. (2020) note that, "with only a few exceptions" the assessments did not evaluate whether managers used the results or whether the application of the results was effective. In contrast, Welch (2026) reports how their approach improved understanding of prioritizing and combining risk of Active Pharmaceutical Ingredients (APIs), leading to better prediction of combined risks of APIs with similar/same mode of action. However, they point out that there was no direct management use (e.g., creation of wastewater treatment plant effluent limitations) and go on to say that their approach is a useful resource for authorities to implement and prioritize translation of EU legislation on wastewater treatment.

Here, we take a broad look at challenges that occur between tool development and incorporation into management decisions to uncover solutions and best practices that will help new tools and approaches become successful. To do this, we consider the utility of a general framework and a system's view to set the context of tool development. Then, based on our experience, that of other authors in this book, and examples from the literature, we compile suggestions on how to overcome barriers to acceptance of new tools. We note how we might define success and what makes some tools successful. Our chapter concludes with a summary of practical advice to help advance best practices. We hope to help spur the conscious determination mentioned by Kapustka (2026) to overcome limitations of legacy approaches and tools that have "stubbornly continued," such as use of NOAECs, LOAECs, and nonscalable and summed risk quotients, and lack of considering stressors other than chemicals such as habitat quality and land use.

This book aspires to advance methods that can address wicked problems such as cumulative risk, and especially to consider newer tools that have great promise and may be underutilized. How to analyze and interpret cumulative risk itself is not new, and there has been long-standing interest in doing so. The problem with advancing this field seems not to lie with the new tools and approaches themselves and whether they can advance the science and produce better informed decisions but rather with lack of a grasp on why they cannot get a foothold in the decision-making process. As discussed by Kapustka (2026), this results in a mismatch between stated protection goals and standard practice methods used to assess them. Langan et al. (2023) have posed and tried to answer the question: "What is necessary to evolve the discussion and mutual understanding between developers and end-users (industry/regulatory agencies) on when and how to further methods/approaches as new science or regulatory change emerges?" Anklam et al. (2022) also note the need for a clear strategy to rapidly evaluate new technologies and determine which are ready for regulatory use because otherwise their value will remain largely unrecognized and unused.

Both examples suggest to us the value of a simple framework from the developer's point of view that considers the context of tools in the larger system so that the evaluation strategy is baked into the process and not added at the end. As Harwell and Jackson (2021) point out, there are many factors other than the scientific/technical limitations that inform decision alternatives, such as budget and time/space constraints, disturbance events, pollutants, changes in adjacent infrastructure, land use, and management. In discussing ecosystem management, Ernstson (2013) has noted that the benefits humans and society derive from ecosystem processes must be viewed as entangled in social and political processes. Lewis et al. (2020) call for including a wider breadth of user groups and social behaviors so that potential interactions, feedback, and trade-offs between groups and uses will be considered in ecosystem management.

It is also important to know the bigger management view and desired outcomes. Even though work may be done on a particular set of stressors and their combined effects, the application, for example, at a contaminated site, may be within a larger goal of remediation that will foster restoration and then economic revitalization. Williams and Hoffman (2020) provide a framework that explicitly addresses the role of humans and recognizes the ability of communities to learn and make choices that improve the environment through translational science. Sample et al. (2023) point out that early in the risk assessment process the problem formulation step should reflect the management needs and incorporate the larger ecological context to "recognize, account for, and to the extent possible, minimize unexpected ecological and human welfare impacts." They state that problem formulation identifies the exposure and toxicological approaches that will address the conceptual model, includes a definition of unacceptable level of impact considering an acceptable level of uncertainty, defines the analysis plan and strategy, and identifies the uncertainties that will exist even after the approach is implemented. Kapustka (2026) argues that landscape perspectives and recognition of the importance of thinking in terms of complex social-ecological systems provides a better foundation for engaging in meaningful dialogue regarding management options to identify and deal with risks. Stahl (2026) gives an example of coordinating remediation and restoration in such a way as to remove a greater mass of contaminated soil and sediment, while providing a much more diverse habitat in the process. The analysis for this coordination was based on projections of potential habitat value increases if restoration were included with the remediation, judged against what could be achieved with the remediation alone.

Lessons from Development and Use of NAMs to Reduce Animal Testing

We selected the area of NAMs for an introductory deep dive due to the high profile (legislative mandates, regulatory connections, and international attention) and concerns on use in regulatory decision-making. The acceptance and use of NAMs encapsulates many of the issues new tools and approaches will face. As Rattner et al. (2023) pointed out, the dissemination of *Toxicity Testing in the 21st Century: A Vision and Strategy* (NRC 2007) accelerated the development and use of in vitro, in chemical, and in silico methods as more efficient, predictive, and ways to inform hazard and risk assessments, but emphasis has been on human health. One NAM in particular, high-throughput (HTP) testing using cell lines, has formed the backbone of the Tox21 Program led by the Environmental Protection Agency (EPA), National Institutes of Health, and the US Food and Drug Administration (FDA). Between 2009 and 2021, the Tox21 Program screened more than 8,500 compounds with over 70 different HTP assays (Richard et al. 2021). The data are available from the USEPA CompTox chemical database (https://comptox.epa.gov/dashboard/; Version 2.2.1 – May 2023; Williams et al. 2017).

According to Thomas et al. (2019) and citations therein, the external community has used CompTox data across a wide range of applications including providing mechanistic support for chemical hazard assessment in pesticide risk assessments (California EPA), screening for substances of high concern (European Chemicals Agency), regulatory and economic impact assessments (European Commission), use as a key component in a tiered toxicity testing framework (USEPA), and identification of potential toxicological liabilities for pharmaceuticals (Pfizer). Per Thomas et al. (2019), Health Canada has also outlined a plan that uses CompTox data in identifying priorities for risk assessment within their Chemicals Management Plan.

Some traction has been gained for use of NAMs in the EPA Toxic Substances Control Act (TSCA) program to prioritize compounds for regulatory action. Initial translation of the developments in CompTox to regulatory decision-making have occurred primarily within the EPA's Endocrine Disruptor Screening Program (Thomas et al. 2019). In 2015, EPA decided to use the high-throughput and computational approaches for predicting estrogen receptor activity to replace specific in vitro and in vivo tests in the Tier 1 testing to prioritize chemicals for further evaluation. The incorporation of NAMs into the EPA TSCA program increased greatly with passage of the Frank R. Lautenberg Chemical Safety for the 21st Century Act in 2016. Further applications of CompTox tools and data have been outlined in the recent release of the strategic plan to promote the development and application of alternative methods in the TSCA program (USEPA, OCSPP 2018) and selecting candidates for prioritization in TSCA. In 2021, EPA released a report (USEPA 2021) that presents an approach that integrates publicly available hazard, exposure, persistence, and bioaccumulation information for chemical substances. This proof-of-concept case study discriminated between high- and low-priority-candidate chemical substances and identified potential information gaps. The report (USEPA 2021) goes on to say that this approach "may be applied to large numbers of chemical substances and is an important tool for efficiently integrating and synthesizing large amounts of publicly available information, and aspects of the approach could be adapted and applied to other prioritization decision contexts."

It is encouraging to see the new methods beginning to be incorporated into regulatory actions. Yet challenges remain for use in decisions. As an example, James et al. (2023) recently combined HTP in vitro assay data with three other biological response measures (Predicted No Effects Concentrations from the European Union NORMAN Ecotoxicology database; hazard concentrations for 5% of species derived from species sensitivity distributions; and screening values from toxicity studies) to prioritize contaminants of emerging concern. Monitoring data from marine water, wastewater, and fish and shellfish tissue samples from 17 different campaigns combined with multiple biological response measures were compared with appropriate threshold values. James et al. (2023) state there are still challenges to be surmounted in using HTP methods. These include extrapolating in vitro data to in vivo impacts, extrapolating from human-based cells and cell components to wildlife, the lack of regulatory frameworks for incorporating in vitro data, and a lack of confidence in the application of new approaches.

Thomas et al. (2019) recognized these issues previously and developed the EPA CompTox blueprint "to obtain broader acceptance of the new approaches and application to higher tier regulatory decisions" (see Figure 11.1 reproduced from Figure 1 in Thomas et al. 2019). These elements address establishing confidence, providing outreach and training, providing the software and tools, and addressing/reducing uncertainty and variability. They are to work in concert with the four areas where HTP is expected to find high use.

Similarly, and more recently, Langan et al. (2023) discussed barriers to adoption of NAMs. They provide activities suggested by various stakeholders that could overcome barriers to use of NAMs (Table 11.1). These include roles for

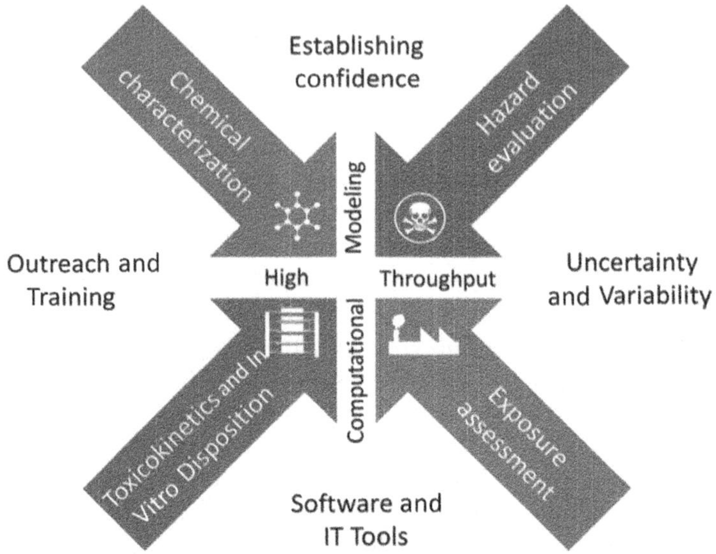

Figure 11.1 Key elements to obtain broader acceptance of the new approaches and application to higher tier regulatory decisions (EPA CompTox blueprint). *Source:* Thomas et al. (2019) / Oxford University Press.

government, industry, the regulatory sector, academia, and journals. We repeat these here because they introduce many of the best practices and suggestions we provide later and illustrate the value of a broad view toward the use of new tools and approaches.

Thomas et al. (2019) state EPA's goal for CompTox is to develop a generalizable and scalable evaluation framework for performance standards. To achieve acceptance in the regulatory community, the framework will provide guidelines and standards to ensure the reproducibility and performance of the methods. To establish scientific confidence, EPA will develop collaborative partnerships with regulators on evaluating alternative methods for regulatory application and performing case studies. EPA will examine a variety of regulatory applications and decision scenarios. The EPA strategy for application of new approach methods to TSCA describes the engagement needed to ensure the methods and data are applicable to regulatory decisions.

Certainly not every new tool or approach has such backing by legislation or government entities. However, building a strategy for tool use in decision-making can benefit from understanding key users and challenges, and one can look for analogies between the challenges NAMs face and those that other tools and approaches under consideration might face.

Table 11.1 Key roles for government, industry, the regulatory sector, academia, and journals in development and acceptance of new approach methodologies.

Government
- Strive for intentional harmonization of NAMs and related legislation
- Increase funding for NAM validation
- Prioritize research that is based on NAMs vs. animal testing
- Formulate policy goals to promote the use of NAMs and shift away from animal testing
- Establish or maintain an international scientific body to support and raise awareness of NAMs among regulators
- Advance the capacity of regulatory toxicology beyond its administrative responsibilities as a scientific field
- Manage the multistakeholder renegotiation needed for the redefinition of safe exposure concentrations when based on NAMs
- Financial support for establishing NAM resources, including infrastructure (e.g., space for cross-disciplinary research outside of academia and independent of typical funding calls), trained personnel (e.g., toxicologists, bio-informaticians, system biologists, animal husbandry/biology experts, analytical chemists), internships for constant recruitment and targeted educational curricula with continuous proofing by academic partners

Industry
- Promote NAM acceptance by the regulatory sector via continuous dialogue during development and application
- Use NAMs in preparing assessment approaches (case studies)
- Commission and submit NAMs information alongside standard information with incentives from government and regulatory agencies to encourage this
- Support academia to develop and refine NAM approaches (e.g., competitions such as the NC3R'S CRACK IT challenges)
- Support regulatory sector in accepting NAMs through continuous dialogue throughout NAM development process and use

Regulatory sector
- Actively engage with the academic sector to deepen understanding of NAMs and offer direction in early development
- Invest in the recruitment and training of NAM experts
- Deepen the understanding of practical limitations and scientific uncertainties of animal testing
- Make the uncertainties in current RAs transparent to familiarize regulatory toxicologists with their current limitations
- Exploit legislative freedoms to use NAM data in a scientific and creative way
- Train "out-of-the-cage" thinking: Before requesting new animal studies according to current legislation, consider how you could regulate if animal testing was no longer allowed? What would need to change?
- Learn from other regulatory sectors and consolidate examples of convergence (e.g., use in vitro ADME data) as with pesticides or make in vitro data the first step in ERA (like human health–based RA)
- Support parallel/safe-harbor approaches – provide incentives for this to compensate the financial impact and alleviate concerns of differences in results to traditional approaches
- Create space and guidance for new methods to quickly be integrated into RA portfolios (dynamic RA)
- Support industry in submitting NAM data through continuous dialogue throughout the NAM development and use process (proactive rather than reactive)

Table 11.1 (Continued)

Academic sector • Build bridges between sectors early in the project to increase social relevance, visibility, and project opportunities, especially for NAM development and characterization • Provide a framework for graduate researchers to synthesize the available literature on a research question in an AOP format relevant to regulation • Review available NAMs for your own area of interest and identify beneficial development of new NAMs • Critically reflect on options for adapting one's research questions to reduce reliance on animal testing • Adopt methods from other laboratories to increase reliability and wider acceptance; outline in resulting publications which NAMs were used or why animals were used if NAMs were available • Publication of registered reports for all studies involving animal experiments, including use of Animal Research: Reporting of in vivo Experiments (ARRIVE) Guidelines (https://arriveguidelines.org) and contribution to study design, with publication guaranteed regardless of outcome. For example, use of the ARRIVE guidelines could be required before a study and submission with any articles that are the result of that study
Journals • Environmental journals could adopt a "Registered Reports" approach to ecotoxicology studies, allowing for transparent publication of study design and results regardless of outcomes • Enforce criteria for reporting on animal studies and NAMs based on AOPs (e.g., ARRIVE criteria) • Insist on transparent reporting of data and methods appended to supplemental information rather than bland instructions to contact the authors • Encourage special issues on AOP development and NAM-based case studies for selected aspects of environmental hazard and risk assessment

ADME, absorption, distribution, metabolism, and excretion; AOP, adverse outcome pathway; ERA, ecological risk assessment; NAM, new approach methodology; RA, risk assessment.
Source: Adapted from Langan et al. (2023); Figure 2.

Framework for Tool Development

Based on our own experiences with national and state regulatory, trustee, and research organizations as well as literature and feedback from authors of other chapters in this book, we find it helpful to structure the lifecycle of tools/approaches from their genesis to use. The flow of integration of new tools can vary, but generally successful integration requires specifying the outcome. The process to develop a guideline informed by the work of FDA's division of applied regulatory science includes the concrete outcome of adopting a guideline (Figure 11.2 below from Chiu et al. 2023). Other processes can be examined such as the EPA IRIS program or the ASTM process for developing standard methods.

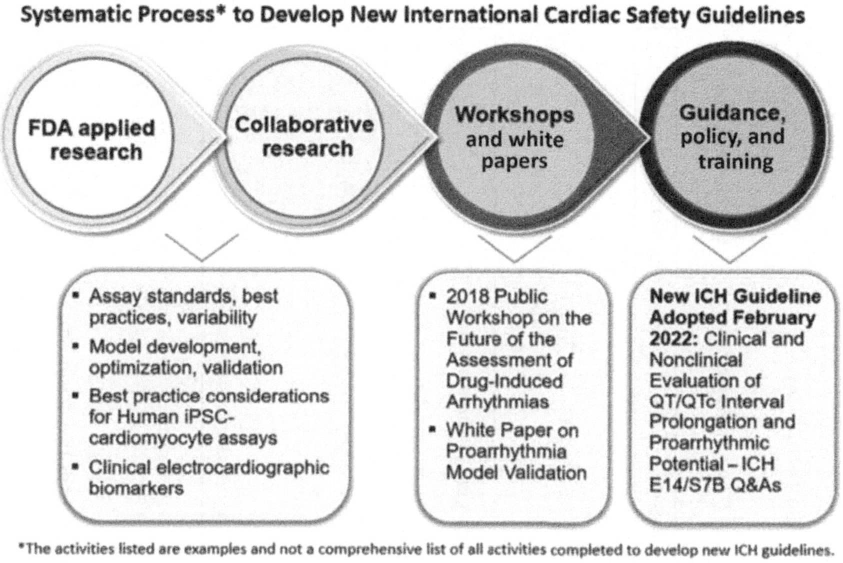

Figure 11.2 Example of systematic process with specified outcome.

We illustrate a more generalized version of what we have experienced and evaluated in identifying key steps from concept to use in decision-making (Figure 11.3). An example is presented that illustrates the steps in the development of an EPA EcoUpdate technical document. This particular example shows the steps to developing guidelines in linking contaminated groundwater discharges to ecological effects in the transition zone between groundwater and surface water (USEPA, OLEM 2008). As indicated (Figure 11.3), the decision-makers were engaged throughout the process and the EcoUpdate provided a framework, decision tree, case studies, and tools for characterizing contaminated discharges as well as potential ecological effects.

While we consider a tool successful when it is used in decision-making, Suter (2023) describes additional aspects of success of innovation in environmental assessment. These include publication in a credible journal or prominent report; publication in a guidance document for an agency or organization; and frequent use in assessments. Setting a goal for success and defining it is important in developing a new tool or approach. Rattner et al. (2023) and Lillicrap et al. (2016) recommend addressing readiness, reliability, reproducibility, relevance, and regulatory acceptance of the new tool which we have included in Figure 11.3. In contrast to the framing presented in Figure 11.3, Clancy et al. (2023), investigating the academic researcher/policymaker interface, confirm that there continues to be a strong bias toward publication as the measure of success in academia but

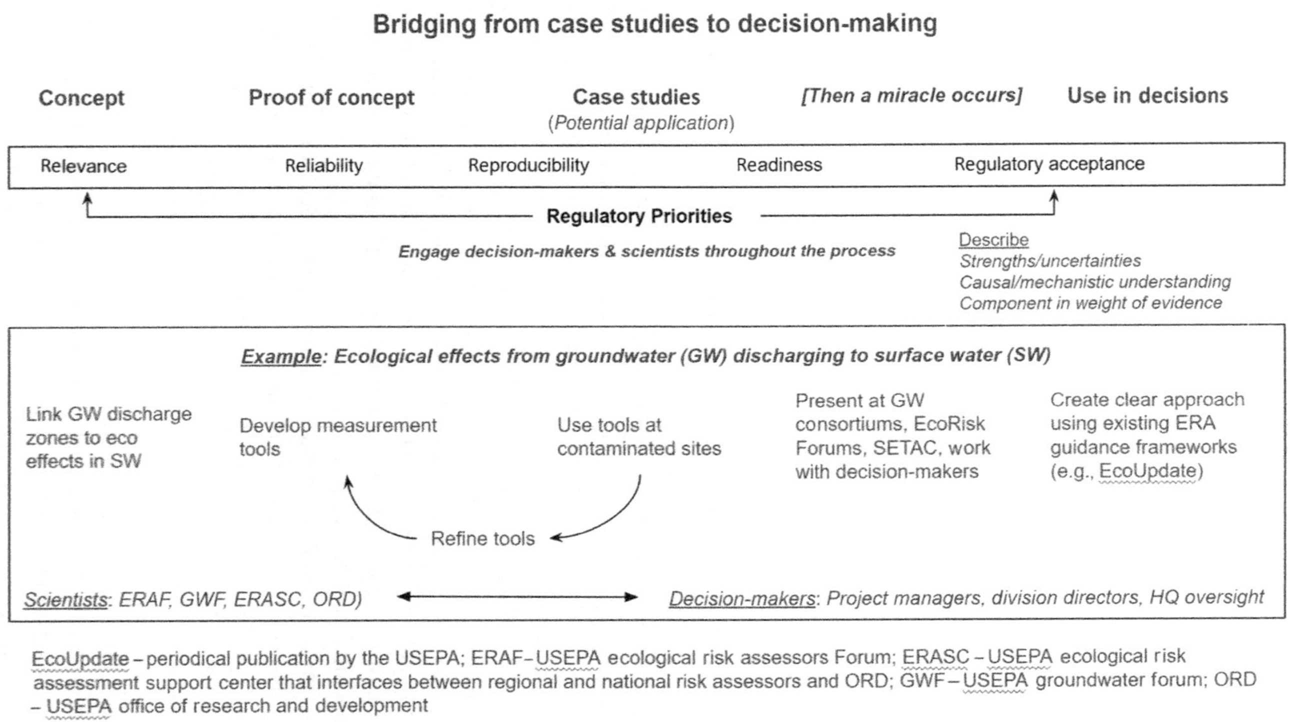

Figure 11.3 Bridging from case studies to decision-making in the use of tools/approaches by decision-makers.

expressed hope that such academic norms could shift to allow academics the time required to build relationships with science funders.

Actions to Advance the Use of New Tools and Approaches

As we considered the issue of resistance to using new tools by decision-makers, we formulated a series of questions based on our practical experience working with decision-makers at multiple levels, from Project Managers at Superfund Sites to Managers at Resource Agencies.

Q1. Why are uses in decisions not easily found?
Q2. How do you unlock obstacles to advancement?
Q3. What creates compelling arguments for decision makers?
Q4. What support for tools is needed for success?
Q5. Where do we see successes?

In addition, we solicited input from other authors in this book (see Appendix) and reviewed challenges, lessons learned, and next steps reported in the literature.

We compiled and categorized the results into suggested actions and included them in the discussion of our five questions. For some actions, we give detailed examples and in-depth discussion; for others that are more self-evident, we provide a succinct description. We cite sources from the literature, other chapter authors (either citing their chapter or as personal communication, if input was response to our query), or our own experience (when no source is given).

Q1. Why Are Uses in Decisions Not Easily Found?

We cannot easily find examples of decisions that relied on the new approaches/tools. Are they reported somewhere that is not obvious to us? For example, the USEPA Office of Research and Development (ORD) bases their workload and allocates resources on a Strategic Research Action Plan (STRAP; USEPA, ORD 2022). A key basis of that plan comes from prioritized National and Regional science needs, identified by decision-makers. The steps in the STRAP clearly link back to delivering products. There is no next step to identify the impact of the products although these are generally internally developed by ORD scientists as part of their opportunities for promotion. Even products of high use and impact such as guidance documents, white papers, etc. often are not as valued as mainstream journal publications. For example, a document on best practices relevant to underground storage tanks has the potential to bring new tools, approaches, analyses, etc. to regulating tens of thousands of underground tanks in each state of the United

States (e.g., Petroleum Vapor Intrusion model; USEPA, ORD 2018). Self-reporting by researchers of uses in decisions is not something commonly encountered in our experience. In EPA's Superfund Program, there is a lot of effort spent to identify high-priority science needs to ORD. And there is clear guidance on developing a defensible Record of Decision for each Superfund Site. Yet one is left to comb through such records to find any evidence of new tools used and how they affected decision-making. This may be easier to do if one has been involved with the site or can identify and contact the Project Manager.

To our knowledge, there is not much marketing that goes on by organizations and agencies to publicize prioritized research needs. Therefore, it can be difficult for a research scientist in the academic sector to find these fruitful intersections and develop collaborations. A notable exception is when science needs are translated to requests for research proposals. However, many of these are aimed at proof of concept or case studies. As we discuss further in this chapter, there is a key role for scientific societies to help bridge this gap as well as for agencies to publish priorities in sufficient detail to make them worth examining for potential collaboration. Can researchers/developers be encouraged to go beyond a vague description of a tool being potentially useful to actually documenting its use as a basis for a decision? One encouraging solution to this that we are familiar with is increased reporting methods such as used by the Strategic Environmental Research and Development Program (SERDP https://serdp-estcp.mil/projects; accessed 3/27/2026). Each project incorporates a tech transfer component to ensure the research or project is shared with entities outside of the program.

Q2. How Do You Unlock Obstacles to Advancement?

It seems reasonable to provide examples and case studies to help move from aspirational to implementable, and many have done so. For example, Welch (2026) notes that, to their knowledge, the object-oriented Bayesian Network (BN) presented is the most in-depth Bayesian model of pharmaceutical environmental risk to date. And they hope it will "serve as a foundation for the eventual development of useful tools for risk assessors and managers."

As others have pointed out, there are actions that can be taken by multiple entities (Langan et al. 2023). We wonder, however, if it is reasonable to expect academic research to unlock bureaucratic paralysis by leading with research examples. Our concern is when research is not clearly linked to decision-making and research success is defined by publications and citations in the primary literature, the link becomes obscure. Can we expect grant funding processes to help advance adoption of new tools when funding is often restricted to basic research? The explicit role of funding is important. For example, if the EPA selects new techniques for use at cleanup sites, they can seek cost recovery. How are academia, consultants, and industry reimbursed for their contribution?

Others have noted a key role for scientific societies, and we concur. The convening of Pellston© workshops that include experts from across government, industry, academia, and consulting has been given as an important example for advancing new methods. We advocate for such workshops to explicitly include decision-making.

To overcome the inertia in the field of environmental toxicology repeatedly identified by other authors in this book (e.g., modern data analysis), we list and briefly discuss several germane topics. Context and systems thinking are important, and many of the following are practical suggestions. Being responsive to decision-makers goes a very long way to gaining their acceptance and use of new approaches.

Q2-Suggested Actions

Understand the Decision Context and Determine Willingness to Use New Approaches

- **Understand the science/policy interface.** As Clancy et al. (2023) state, "new facts alone cannot directly inform policy." Informing policy requires the ability to understand and interface with science policy makers.
- **Lay out the context, assessment objectives and criteria, and roles of those involved.** This is fundamental to allow for the take up of CEA outcomes in management processes (Stelzenmüller et al. 2020).
- **Recognize that complexity, time, cost, and data requirements are seen as barriers by managers** (Lewis et al. 2020).
- **Recognize the constraints on users of assessments** (Suter 2023).
- **Recognize that regulated entities will be reluctant to use new approaches.** This is when project managers are most comfortable with conventional approaches and unlikely to use a new approach (Rattner et al. 2023). Regarding application of NAMs, managers are "often most comfortable with hazard quotient approaches, and few responsible parties are likely to invest in a NAM-based ERA in the absence of an expectation that the new method will be considered by the remedial project manager (RPM)."
- **Determine the questions policy makers most need answering by spending time at other institutions.** Clancy et al. (2023) suggest academics spend time in government service positions to build relationships with the institutions their research seeks to inform and to learn about the institution's goals, constraints, and objectives. They also call for government institutions to create more part-time or short-term opportunities to build relationships with academics. A good example of this is the Oak Ridge Institute for Science and Education (ORISE) program (ORISE 2023), which provides graduate students and postdocs the opportunity to work in a federal agency. And many federal agencies

allow staff to be on a detailed assignment to other government agencies. Clancy et al. (2023) further suggest that academics can benefit greatly from working with institutions such as think tanks that translate academic research into social impact especially when those institutions facilitate engagement among federal science agency staff, academics, and outside practitioners who may be struggling to add that engagement on top of their regular jobs.

- **Find science priorities and connect with end users.** The first aspect of relevance involves determining who the end user is and what decisions they need to make (Figure 11.1). Published lists of research priorities such as those provided by the European Medicines Agency (EMA 2023) can be important drivers to developing new tools and their adoption because the end uses are clearly defined. The value of matchmaking has recently been mentioned. Clancy et al. (2023) give the example of the Sabbaticals in Service program that enables matches between academics who are interested in working with federal agencies and agencies that are searching for experts in particular areas. Understanding end-user priorities is extremely important to the success of new tools. For example, the USEPA ORD periodically publishes its Strategic Research Action Plan. One part of this plan, the Sustainable and Healthy Communities program (USEPA 2022), resulted from negotiations between ORD, EPA National Programs, and EPA Regional Programs. This approach helps establish collaborations and partnerships between developers and end users. The 2022 plan addresses three broad research topics: (1) Advancing Remediation and Restoration of Contaminated Sites (including, among other subtopics, Site Characterization and Remediation; Chemicals of Emerging and Immediate Concern); (2) Materials Management and Beneficial Reuse of Waste (including Landfill Management Under Climate Change; Waste Recovery and Beneficial Use of Materials); and (3) Integrated Systems Approach to Building Healthy and Resilient Communities (cumulative impacts of stressors and exposures, especially in overburdened and underresourced communities). Under these topics, one can find the development of specific products. These are tangible scientific or technical deliverables that address the research needs of ORD and its partners. Research Area Coordination Teams, made up of ORD scientists and engineers, EPA program and regional staff, and state members, collaborate to determine the individual research products corresponding to each output. When agencies publish their science priorities in this kind of fashion, it facilitates the development of partnerships and connections with agency scientists.
- **Determine the regulatory priority of the context/use of the tool.** For example, Welch (2026) determined that the regulatory priority of pharmaceutical ERAs was low.
- **Fill a clear need** (Suter 2023).

- **Make ecological methods analogous to human health methods.** Given the priority of human health risks and effects and the familiarity with those methods, it can be effective to show how your method is analogous (Suter 2023).
- **Provide and publish counterexamples with new approaches.** Be prepared to encounter inertia in reviewers of new methods (W. Landis, Personal Communication).
- **Develop people skills.** Stahl (2026) states that one of the most important learnings from his career was the need to develop people skills, which is generally not taught in graduate school or during a postdoctoral fellowship. Even the most technically competent individual will have difficulties in their career if they cannot work and communicate well with others.

Build Confidence in New Approaches by Showing the Value to Risk Assessors

- **Provide clear technical basis and practical guidance**. Greenberg et al. (2014) acknowledge that regulatory applications of passive sampling methods (PSMs) have been limited but felt they were providing the technical basis and practical guidance needed to justify increased application.
- **Meet the requirements of risk assessors**. To do this, identify the decisions that the ERA is intended to inform. Identify the metrics identified in problem formulation that are needed to inform the decision. Determine the quality, quantity, and uncertainty that are acceptable for the decisions. Link to exposure measures that will be used (i.e., effects need to match measures of exposure to be integrated into the ERA; Rattner et al. 2023).
- **Match the model/tool to the scale of the goal or question.** Consider several levels of complexity for the tool, for example, a simple version that can be used for an initial screening analysis may be appropriate (Lewis et al. 2020).
- **Communicate uncertainty to avoid misinterpretation of results** (Lewis et al. 2020).
- **Demonstrate the value for regulatory approaches that retain critical information on the different types of uncertainties**. Keep as much information as possible versus condensing information into an index or even simple distribution (W. Landis, Personal Communication).
- **Demonstrate successful applications and case studies at conferences, meetings, webinars, and other outlets** (Greenberg et al. 2014).

Bring in Stakeholders Early and Throughout the Process

- **To improve relevance and usability, resource managers and other stakeholders must be part of the model development process to identify important decision metrics** (Lewis et al. 2020).
- **Engage stakeholders early.** Early engagement will help to increase understanding of new tools and promote acceptance and limitations of their use

(Greenberg et al. 2014). For example, at contaminated sites, Stahl (2026) insists to set clear remediation management goals and conduct frequent outreach to stakeholders inside and outside of your organization because no "one size fits all" with respect to remediation.

- **Ensure communication is friendly to public dissemination** (M. Ottinger, Personal Communication).

Q3. What Creates Compelling Arguments for Decision-Makers?

Demonstrating success with a new technology or process is necessary but not sufficient. For example, there has been a huge pushback on use of BNs as a framework to address mixtures despite keen interest and conducting many successful case studies around the world (W. Landis, Personal Communication). One needs to be aware that scientists and decision-makers are human and prone to bias based on political and economic interests, and that bias may lead to consensus not based on evidence (Socol et al. 2019). If decision-makers are looking for scientific consensus, Socol et al. (2019) caution that there are examples of decades-long scientific consensus on erroneous hypotheses and propose, very interestingly, that, for policy advice purposes, a scientific statement or model should be considered as the subject of proper scientific consensus only if "shared by those who would directly benefit from proving it wrong."

In our experience, decision-makers generally want a clear benefit relative to cost. They see new tools, especially complex ones, as requiring staff time and staff training or the need to acquire funding to hire consultant expertise. And external consultant input still leaves the manager needing a review of the results by in-house or in-agency experts. Managers want to see minimal risk of failure, clear improvement, and demonstrated impact/utility. On top of this, many managers are reluctant to be the first to use a new approach in case their decision is challenged. Thus, they look to sanctioned guidance.

Inroads can be made with managers by understanding the "newness" from their point of view. What is the utility of the new tool/approach? Does it address a priority? Can it create the results within a reasonable time frame? Have managers been engaged early on so it addresses their issues and constraints? We have seen success with managers when we engage them early, even at proof of concept, recognize their priorities and timelines, and crosswalk between the new and established methods, showing how the new approach fits within existing guidance. Managers may be more amenable to a new approach if they are dealing with a potentially unique situation where modifications are necessary (e.g., use of drones and hyperspectral analysis to evaluate metals on steep or unstable slopes that pose risk for field investigators).

Q3-Suggested Actions

Value of Frameworks

- **Use structured approaches.** These are useful to organize a stepwise process to address complex challenges and reassess approaches to adjust as needed, such as with adaptive management (M. Ottinger, Personal Communication).
- **Show how new tools can be incorporated into existing frameworks.** For example, Lewis et al. (2020) suggest using the DPSIR (Driver, Pressure, State, Impact, Response) conceptual framework to improve adoption of ecosystem goods and services (EGS) models in coastal and marine planning. Starting in 2015, the European interdisciplinary research project AQUACROSS developed an assessment framework for Europe's aquatic ecosystems and applied it in eight case studies (https://www.aquacross.eu/). Lewis et al. (2020) noted that as more studies demonstrate the utility of EGS modeling within ecosystem-based management situations, knowledge and experience from these cases can be used as reference to further develop these frameworks (M. Ottinger, Personal Communication).
- **Develop a standardized framework for model implementation.** For example, this is recommended by Lewis et al. (2020) to help adoption of ecosystem goods and services models into coastal and marine planning processes.

Model/Tool Selection

- **Provide model selection tools that link to management alternatives.** The best model will be the one most appropriate for the situation, data, and question at hand and define the most practical management alternatives to test (Lewis et al. 2020).
- **Develop tiered or coupled models to include multiple drivers and cumulative impacts** (Lewis et al. 2020).
- **Show how the tool is usable.** Make tools field-friendly and convince field biologists to use them (M. Ottinger, Personal Communication).

Show the Value for Decision-Makers

- **Take the time to present evidence to overcome skepticism.** Have enough time and evidence to show that the new tool or approach is not a way to "hide the ball" (R. Stahl, Jr. Personal Communication). For example, when facing skepticism on using a multiple-stressor approach that might suggest the chemical(s) in question were not the greatest source of ecological risk (R. Stahl, Jr. Personal Communication).

- **Evaluate the effects of not achieving objectives.** For example, a Cumulative Effect Assessment should be founded on established risk criteria reflecting the selected policy objectives to maximize the advice the assessment can deliver in the management process (Stelzenmüller et al. 2020).
- **Show how the inclusion of complexity and the influence of other variables make models more informative and desirable for decision-making** (Lewis et al. 2020). This could help overcome the lack of acceptance of spatially explicit models that account for complex behaviors and spatial movements that were published more than a dozen years ago (Sample et al. 2023). One way is to demonstrate how complex models inform adaptive management. Landis (2026) states that ecological risk assessment can inform adaptive management of natural resources as well as mitigation strategies (Landis et al. 2017). An example of a complex model evaluating the influence of multiple stressors is the Bayesian Network–Regional Risk Model (BN-RRM). This model identifies relative risk to various endpoints, identifies which stressors are most important in influencing that risk, and uses counterfactuals to calculate management goals for ecological variables. This helps decision-makers to prioritize stressors for mitigation efforts and to determine the goals of those mitigation efforts. The BN-RRM allows for the inclusion of new data as they become available, providing new information to decision makers on progress of mitigation efforts as well as changing environmental factors. This iterative approach to ecological risk assessment and management is critical for uncovering unforeseen effects and explicitly addressing uncertainty (e.g., under climate change).
- **Produce measurable predictions.** For instance, Landis et al. (2017) were able to relate various stressors to impacts on various politically important endpoints (human health and water use, population of key species), which had the added benefit of producing more measurable predictions that can be subsequently observed in real-world conditions.
- **Include the complexities of layers of stressors and factors that exacerbate the potential for adverse effects** (M. Ottinger, Personal Communication).
- **Broaden your focus to include nonchemical stressors.** Too singular a focus on chemicals has hindered the inclusion of climate change model projections in ERA. Recent manuscripts submitted to the journal Integrated Environmental Assessment and Management show clearly that some elements of climate change model projections can be incorporated into ecological risk assessments and at spatial scales appropriate for making risk management/climate adaptation decisions (R. Stahl, Jr. Personal Communication).
- **Keep the method as simple and easily understood as the problem allows** (Suter 2023).

- **Clearly identify the linkage between risk conclusions and risk-management goals and embrace a more systems view** (Sample et al. 2023).
- **Collect enough of the right data to answer the question.** Consider how data collection using new methods will affect decisions, given all possibilities, and develop a decision plan a priori (Sample et al. 2023).
- **Refrain from adopting arcane/exotic/obscure techniques** (Suter 2023).
- **Show the value of causal assessments vis-à-vis risk assessment** (Suter 2023).
- **Promote the use of cause–effect conceptual models with exposure response modeling and specific endpoints being applied** (W. Landis, Personal Communication).
- **Determine the level of risk of adverse effects that managers will tolerate**. The tolerated risk should be reflected in risk criteria, such as definitions of effect sizes (e.g., defining the degree of change of an ecosystem state due to a certain amount of exposure) or thresholds in relation to acceptable levels of pressures remaining within management boundaries (Stelzenmüller et al. 2020).
- **Demonstrate successful applications and case studies at conferences, meetings, webinars, and other outlets** (Greenberg et al. 2014).
- **Be prepared for a lot of explanations and examples.** For example, contributing results from a new application (e.g., BN analysis) to the decision-making process took a lot of explanation and examples to demonstrate how the process works (W. Landis, Personal Communication).
- **Include processes that incorporate decision-making.** For example, when planning data collection and use, use a quality assurance plan that incorporates performance standards linked to decision-making.

Q4. What Support for Tools Is Needed for Success?

Determine the level of support the tool/approach will need. In many cases, having a champion at high levels in your organization or that of potential users is extremely helpful. Include a support plan in your overall approach and consider developing guidance, technical support, and providing a means for a burgeoning cadre of end users to interact.

Q4-Suggested Actions

Technical Basis and Regulatory Guidance

- **Regulatory authorities should develop and issue guidance documents and operating manuals** (Greenberg et al. 2014).
- **Use workshops.** For example, regarding the evaluation of climate change along with chemical stressors, R. Stahl, Jr. (Personal Communication) expressed the

hope that manuscripts from workshops will dislodge ecological risk assessors and decision-makers from their singular focus on chemicals. Stahl (2026) gives the example of how specific risk assessment topics have been addressed through the SETAC Pellston© Workshop program.

- **Work with and present to Science Advisory Boards**. Science Advisory Boards influence policies relating to new methods. However, see recommendations for transparency in constructing and monitoring such boards (Bipartisan Policy Center 2009).

Assistance

- **Describe how users can obtain assistance especially for early adoption of the new tool/approach.** For example, when the USEPA ORD scientists develop new tools, they generally provide initial assistance for proof of concept and case studies. After that they assist with novel or challenging applications to build out a more inclusive portfolio of case studies. In some cases, extensive assistance is provided (e.g., CompTox program; Williams et al. 2017).
- **Recognize that oversight will be needed for new tools.** Even if consultants and industry practitioners learn new advanced techniques, there is the need for regulatory oversight, which will require resources and personnel as well as training in modeling and programming or other skills needed to use the tool. Trusted consultants with such skills are likely to be needed.

Develop Expertise to Support the New Methods

- **Increase the data availability and expertise.** For example, data and expertise are needed for running complex models such as Individual-Based Models (Lewis et al. 2020).
- **Overcome limitations of datasets not designed for the decision at hand.** For example, Welch (2026) described how the big dataset they accessed was not designed for the questions they were interested in. They evaluated mixture toxicity of a broad range of active pharmaceutical ingredients with similar receptor targets but encountered inconsistent data availability across groupings.
- **Commit to training practitioners in the use of new tools** (Greenberg et al. 2014). Rattner et al. (2023) give an example of models that link organism-level responses to population or higher-level ones. Use of these models is hindered by their complexity, uncertainty, resource investment, and data availability. Additionally, practitioners must be trained in the nuances of population or ecosystem-level modeling, which requires a commitment to studying the field and developing strong modeling and programming skills. For example, Margiotta-Casaluci et al. (2023) recently reviewed the need for extrapolating biological data to determine environmental safety of pharmaceuticals. They recognize that the translation of comparative ecotoxicology research into

real-world applications depends on a limited availability of skilled practitioners and so call for efforts to "support and strengthen comparative toxicology research and education at a global level." This is echoed by Landis (Personal Communication), who states that new ecological risk assessors need a solid foundation in data analysis and experimental design that includes modern tools such as those used to describe exposure–response.

- **Include certification of training.** Provide data analysis training with a certificate for seniors and grad students that covers newer methods to analyze exposure–response data (W. Landis, Personal Communication).

Validate the New Methods

- **Validate methods.** For example, to use adverse outcome pathways in wildlife ecological risk assessment, Rattner et al. (2023) point to the need to establish quantitative pathways to higher-level effects (currently most are qualitative). They state that validation must cover exposure–response relationships, specificity to a chemical or group, links/relevance to apical endpoints (higher-level effects) and the uncertainties (accuracy, precision, and reproducibility) around these links, and influence of nonchemical stressors on the method. For example, lead impairing behavior in mallard ducklings may not be considered in current ERAs as compared with hematology and biochemistry values. Inclusion would require "quantitative evidence linking impaired learning and memory in a representative organism to (1) alterations in behavior, (2) reduced productivity or survival, and (3) reduced populations."

Q5. Where Do We See Successes?

We provide a few examples to illustrate the value of looking at what contributes to the success of new tools and approaches (e.g., new technologies that are more flexible, reliable, cost-effective, and can work easily with "gold" standard approaches).

Q5-Suggested Actions

Technology

- **Real-time sensors.** Improvements in new field-level instruments are examples of consistent success. One classic example in water quality measurements is the shift from measuring dissolved oxygen (DO) with a calomel electrode to using optical probes. Interface-based sensors are an established technology for the marine environment, with sensors such as the DO optode, optical backscattering, and Chlorophyll-a sensors in use (Briciu-Burghina et al. 2023). Future developments are likely to include sensors for algal toxins, key groups of bacteria, and other organisms through enumeration of DNA or RNA sequences. Molecular biology techniques require small sample and reagent volumes (microliter scale), which makes them ideal for lab-on-chip (LOC) integration. Innovation is becoming commonplace due to recent advancements in microfluidics and LOC devices (Briciu-Burghina et al. 2023).

- **eDNA.** The US Forest Service (USFS) uses environmental DNA to determine presence and potentially abundance of cold-water fish species. Collection is generally an easy field technique (Carim et al. 2016), and results are helpful in understanding if target species are present, which is especially important for Endangered Species Act and listed species. In addition, samples can be collected by citizens, new advancements can provide information on relative abundance, and more species barcodes for reading the DNA results are coming online. Metabolomics is one field where eDNA is used extensively to determine which microbes are present, the functions they perform, and how the functions change in response to stress (Sun et al. 2019).
- **Cross-species extrapolation.** As an example, the Interspecies-Correlation-Estimation (ICE) model (Raimondo et al. 2007) was developed to improve Ecological Risk Assessments for threatened or endangered species. It has been applied primarily to aquatic ERAs. Models using ICE are available for predicting acute toxicity values for birds and mammals but are less robust than for aquatic species. Another example is the Sequence Alignment to Predict Across Species Susceptibility (SeqAPASS; [LaLone et al. 2016]) tool, which compares protein-sequence similarity across species for conservation of known chemical targets, providing an initial line of evidence for extrapolation of toxicity knowledge (LaLone et al. 2018, 2023). Case studies have been developed for targets of interest to the US Endocrine Disruptor Screening Program and the Organisation for Economic Cooperation and Development. The results have been used to identify organisms for suitable follow-up toxicity testing. Rattner et al. (2023) have suggested that tools like SeqAPASS are relevant for wildlife and speculate that as more wildlife-specific tools emerge, molecular and cellular endpoints will likely become common in monitoring programs to detect or predict adverse effects in populations. Recently, Cheng et al. (2021) used SeqAPASS with molecular modeling workflow to inform PFAS bioaccumulation potential across species. SeqAPASS predicted similar PFAS bioaccumulation potential between humans and most vertebrates. In conducting chemical safety evaluations, Mayasich et al. (2023) applied the tool to understand chemical–protein interactions and their effects across species. Schaupp et al. (2023) used SeqAPASS to derive provisional values to guide chemical prioritization as a complement to traditional Aquatic Life Criterion derivation. LaLone et al. (2021) have called for the formation of a global cross-sector collaborative consortium aimed to advance the development and implementation of cross-species extrapolation methods in regulatory toxicology. The authors have led the formation and launch of a new International

Consortium to Advance Cross-Species Extrapolation in Regulation (ICACSER; https://www.setac.org/page/scixspecies).

Processes Using New Methods

- **Crosswalk new methods with the existing ones.** While many reports have identified biochemical, physiologic, and histopathologic biomarkers of contaminant exposures, such nonstandard endpoints have rarely been linked to higher-order effects on wildlife populations (Rattner et al. 2023). Rattner et al. (2023) cite the case of eggshell thinning where laboratory and field studies linked nonstandard endpoints to population-level effects. They also include the example of evaluating avian embryo toxicity from dioxin-like compounds (DLCs) based on the binding characteristics that has allowed three avian groups to be developed based on differences in sensitivity to DLCs.
- **Recognize that stressors will change.** Stahl (2026) noted that frequently the suite of stressors will change during an ecological risk assessment, requiring modifications to the risk-based decision-making process to address site-specific conditions.
- **Move from expert judgment to systematic approaches to address uncertainties.** Thomas et al. (2019) in discussing computational toxicology note that the reliance on expert judgment to address uncertainties has hindered its use for regulatory decision-making. They state that the benefits of systematic approaches for read-across include quantifying the uncertainty and providing a benchmark to assess whether other contexts of similarity (e.g., physicochemical, metabolic, or biological as assessed using HTP Screening data) reduce the uncertainty.
- **Create quantitative linkages of nonstandard endpoints to standard endpoints.** One example from Rattner et al. (2023) is the linkage from molecular endpoints to determinants of population status (such as rates of survival and reproduction) as well as to interactions at the community and ecosystem levels. Endpoints at many levels of biological organization have the potential to improve efficiency, reliability, and realism for the longer term. The value of new approach methods to ecotoxicological hazard assessment has been acknowledged for some time, but applications for terrestrial wildlife are less apparent (Rattner et al. 2023). While food availability has long been acknowledged as a regulating factor of wildlife populations and community dynamics, its consideration in contaminant-driven ERAs is rare because studies are limited.
- **Resolve the issues of extrapolating organism-level effects to population-level analysis.** Sample et al. (2023) state this extrapolation should include differences in organism and life-history characteristics.
- **Align tools with established practices and clarify their integration.** Resource agency publications aligning tools with established practices and

clarifying their integration are a useful step in promoting tool use and bridging from case studies to decision-making. For example, the USEPA Engineering Forum published a a framework for using Ecosystem Services at contaminated sites that aligns potential uses with the ERA paradigm and risk remediation (USEPA Engineering Forum Issue Paper 2017),

- **Develop clear risk criteria for a new approach.** For example, Stelzenmüller et al. (2020) state such criteria should be developed before initiating a cumulative effect assessment, and this should be done within the context and the scope of relevant policies and in consultation with stakeholders. In discussing integration and use of PSMs, Greenberg et al. (2014) call for making key information on their applicability to sediment sites accessible to regulators, responsible parties, and others. They further state that a bridge from research and development literature on PSMs to practical applications of PSMs at sediment sites for potential users must be developed.
- **Be aware of what others are doing.** Awareness should extend to other countries, States, other resource agencies, and approaches. For example, crosswalk new methods from damage assessment, endangered species assessment, time to extinction approaches, etc. into ecological risk assessment.
- **Find where new approaches are being used and seek out key players.** For example, BN approaches have been the norm in the USFS and have been used for wildlife management. In addition, fisheries management in Finland has been done using BNs for some years, but that is because of a few key players. In both cases, there was a small group of people that advanced the ideas along with a small group that adopted them (W. Landis, Personal Communication).
- **Learn from other fields.** Look to other fields such as synthetic biology and molecular worlds to determine why they are quicker to adopt change (W. Landis, Personal Communication).
- **Look for analogies across ecosystem components.** Those with focus in a different field or specialty such as terrestrial, aquatic, wildlife, parasitism, etc. may be developing tools and approaches that can be easily modified and imported.

Communication

- **When creating project summaries, highlight how results were used in decision-making.** For example, SERDP posts project summaries and highlights (https://serdp-estcp.mil/projects; accessed 3/27/2026).
- **Report on use in decision-making whenever possible.** Duncan et al. (2007) provided an example of reporting the management decision for a study of subtidal discharge of contaminated groundwater to a waterway. The results of the study led to development of a scope of work to characterize the nature and extent of subtidal contamination and groundwater flux and were incorporated

into a Sampling and Analysis Plan. At the time of submission of the 2007 paper, several of the initial and step-out exploratory borings had already been completed (Anchor QEA, LLC 2016). See also the earlier discussion on NAMs and Thomas et al. (2019) for more examples.

Summary and Conclusions

Much has been discussed thus far about what to consider and possible actions that can improve the likelihood a novel tool or approach will be used in decision-making. In retrospect, the following stand out regarding barriers and how to overcome them. Perhaps the most important things to recognize are as follows:

Process. Develop a plan that charts the progress of a tool and in so doing explicitly include the other items below. This is akin to creating a conceptual model in a risk assessment, even if your own part in this plan is a smaller slice, such as developing a case study. Select a study that will resonate with your network and be technically convincing.

Network. Make sure to include policy authorities and key researchers, especially those interested in the application of the tool. Policy authorities are likely to be more comfortable if your tool is part of a guidance document or some other approved vetting process such that it may then be used with confidence and be considered defensible.

Technical Rigor. This may be your most comfortable space, but beware just assuming that the technical improvement will be obvious. Anticipate the questions end users will have; better yet, gather those as you build your network. End users are likely to want to know a lot about why your tool is better. They may ask: Does it reduce uncertainty and/or type 1 or 2 errors? Is it better for site-specific applications such as bioavailability? Is it protective as well as realistic and environmentally relevant? Is it defensible in court? Have you cross-walked it with approaches that have policy or regulatory approval? There are many suggestions presented previously about presenting your tool to make your scientific case. Keep finding successes and understanding the elements that led to their success.

Interpersonal skills. You will encounter skepticism, unease, etc. Be prepared to not take this personally but see it as an opportunity to understand and clarify specific barriers such as lack of familiarity, lack of policies supporting the tool, and then see what adjustments or additions you can make to increase the value to potential end users. When conducting research, be open to adjustments or additions you can make to increase the value to potential end users.

Checklist of Actions

These actions have been recommended for improving the likelihood that new tools and approaches will be used in decision-making. Not all will apply, but all are worth considering.

Understand the decision context and determine willingness to use new approaches.

Context:

1) Understand the science/policy interface.
2) Lay out the context, assessment objectives and criteria, and roles of those involved.
3) Recognize that complexity, time, cost, and data requirements are seen as barriers by managers.
4) Recognize the constraints on users of assessments.
5) Recognize that regulated entities will be reluctant to use new approaches.

Approaches for Developing Your Ideas and Tools:

6) Determine the questions policy makers most need answering by spending time at other institutions.
7) Find science priorities and connect with end users.
8) Determine the regulatory priority of the context/use of the tool.
9) Fill a clear need.
10) Make ecological methods analogous to human health methods.
11) Provide and publish counterexamples with new approaches.

Skills:

12) Develop people skills.

Build confidence in new approaches by showing the value to risk assessors.

13) Provide clear technical basis and practical guidance.
14) Meet the requirements of risk assessors.
15) Match the model/tool to the scale of the goal or question.
16) Communicate uncertainty to avoid misinterpretation of results.
17) Demonstrate the value for regulatory approaches that retain critical information on the different types of uncertainties.
18) Demonstrate successful applications and case studies at conferences, meetings, webinars, and other outlets.

Bring in stakeholders early and throughout the Process.

19) To improve relevance and usability, resource managers and other stakeholders must be part of the model development process to identify important decision metrics.

20) Engage stakeholders early.
21) Ensure communications are friendly to public dissemination.

Value of Frameworks:

22) Use structured approaches.
23) Show how new tools can be incorporated into existing frameworks.
24) Develop a standardized framework for model implementation.

Model/Tool Selection:

25) Provide model selection tools that link to management alternatives.
26) Develop tiered or coupled models to include multiple drivers and cumulative impacts.
27) Show how the tool is usable.

Show the value for decision-makers.

28) Take the time to present evidence to overcome skepticism.
29) Evaluate the effects of not achieving objectives.
30) Show how including complexity and the influence of other variables make models more informative and desirable for decision-making.
31) Produce measurable predictions.
32) Include the complexities of layers of stressors and factors that exacerbate the potential for adverse effects.
33) Broaden your focus to include nonchemical stressors.
34) Keep the method as simple and easily understood as the problem allows.
35) Clearly identify the linkage between risk conclusions and risk-management goals and embrace a more systems view.
36) Collect enough of the right data to answer the question.
37) Refrain from adopting arcane/exotic/obscure techniques.
38) Show the value of causal assessments vis-à-vis risk assessments.
39) Promote the use of cause–effect conceptual models with exposure response modeling and specific endpoints being applied.
40) Determine the level of risk of adverse effects that managers will tolerate.
41) Demonstrate successful applications and case studies at conferences, meetings, webinars, and other outlets.
42) Be prepared for a lot of explanations and examples.
43) Include processes that incorporate decision-making.

Technical Basis and Regulatory Guidance:

44) Regulatory authorities should develop and issue guidance documents and operating manuals.
45) Use workshops.
46) Work with and present to Science Advisory Boards.

Assistance:

47) Describe how users can obtain assistance especially for early adoption of the new tool/approach.
48) Recognize that oversight will be needed for new tools.

Develop Expertise to Support the New Methods:

49) Increase the data availability and expertise.
50) Overcome limitations of datasets not designed for the particular decision at hand.
51) Commit to training practitioners in the use of new tools.
52) Include certification of training.

Validate the New Methods:

53) Validate methods.

Technology:

54) Real-time sensors.
55) eDNA.
56) Cross-species extrapolation.

Processes Using New Methods:

57) Crosswalk new methods with the existing ones.
58) Recognize that stressors will change.
59) Move from expert judgment to systematic approaches to address uncertainties.
60) Create quantitative linkages of nonstandard endpoints to standard endpoints.
61) Resolve the issues of extrapolating organism-level effects to population-level analysis.
62) Align tools with established practices and clarify their integration.
63) Be aware of what others are doing.
64) Find where new approaches are being used and seek out key players.
65) Learn from other fields.
66) Look for analogies across ecosystem components.

Communication:

67) When creating project summaries, highlight how results were used in decision-making.
68) Report on use in decision-making whenever possible.

References

Anchor QEA, LLC (2016). Occidental chemical corporation Tacoma groundwater site data summary report. https://apps.ecology.wa.gov/cleanupsearch/document/61620 (accessed 27 March 2026).

Anklam, E., Bahl, M.I., Ball, R. et al. (2022). Emerging technologies and their impact on regulatory science. *Exp. Biol. Med.* 247 (1): 1–75. https://doi.org/10.1177/15353702211052280.

Bipartisan Policy Center (2009). *Science for Policy Project: Improving the Use of Science in Regulatory Policy (Final Report, August 5, 2009)*. Washington, DC: Bipartisan Policy Center. https://bipartisanpolicy.org/download/?file=/wp-content/uploads/2019/03/BPC-Science-Report-fnl.pdf.

Briciu-Burghina, C., Power, S., Delgado, A., and Regan, F. (2023). Sensors for coastal and ocean monitoring. *Annu. Rev. Anal. Chem.* 16 (1): 451–469. https://doi.org/10.1146/annurev-anchem-091922-085746.

Carim, K.J., McKelvey, K.S., Young, M.K. et al. (2016). *A Protocol for Collecting Environmental DNA Samples from Streams*. Gen. Tech. Rep. RMRS-GTR-355, 18. Fort Collins, CO: U.S. Department of Agriculture, Forest Service, Rocky Mountain Research Station. https://doi.org/10.2737/RMRS-GTR-355.

Cheng, W., Doering, J.A., LaLone, C., and Ng, C. (2021). Integrative computational approaches to inform relative bioaccumulation potential of per- and polyfluoroalkyl substances across species. *Toxicol. Sci.* 180 (2): 212–223. https://doi.org/10.1093/toxsci/kfab004.

Chiu, K., Racz, R., Burkhart, K. et al. (2023). New science, drug regulation, and emergent public health issues: the work of FDA's division of applied regulatory science. *Front. Med.* 9: https://www.frontiersin.org/articles/10.3389/fmed.2022.1109541.

Clancy, M., Correa, D., Dworkin, J. et al. (2023). Want to speed up scientific progress? First understand how science policy works. *Nature* 620 (7975): 724–726. https://doi.org/10.1038/d41586-023-02602-9.

Duncan, P.B., Greenberg, M.S., Leja, S. et al. (2007). Case study of contaminated groundwater discharge: how in situ tools link an evolving conceptual site model with management decisions. *Integr. Environ. Assess. Manag.* 3 (2): 279–289. https://doi.org/10.1897/IEAM_2006-039.1.

EMA (2023). Advancing regulatory science in the EU – mid-point report published. European Medicines Agency. 22 March 2023. https://www.ema.europa.eu/en/news/advancing-regulatory-science-eu-mid-point-report-published (accessed 27 March 2026).

Ernstson, H. (2013). The social production of ecosystem services: a framework for studying environmental justice and ecological complexity in urbanized landscapes.

Landsc. Urban Plann. 109 (1): 7–17. https://doi.org/10.1016/j.landurbplan.2012.10.005.

Greenberg, M.S., Chapman, P.M., Allan, I.J. et al. (2014). Passive sampling methods for contaminated sediments: risk assessment and management. *Integr. Environ. Assess. Manag.* 10 (2): 224–236. https://doi.org/10.1002/ieam.1511.

Harwell, M.C. and Jackson, C.A. (2021). Synthesis of two decades of USEPA's ecosystem services research to inform environmental, community, and sustainability decision making. *Sustainability* 13 (15): 1–8249. https://doi.org/10.3390/su13158249.

James, C.A., Sofield, R., Faber, M. et al. (2023). The screening and prioritization of contaminants of emerging concern in the marine environment based on multiple biological response measures. *Sci. Total Environ.* 886 (August): 163712. https://doi.org/10.1016/j.scitotenv.2023.163712.

Kapustka, L.A. (2026). Chapter 2: Foundations of environmental science. In: *Solving Complex Environmental Challenges from Multiple Stressors* (ed. L.A. Kapustka, W.G. Landis, D.J. Ostrach, and C.A. Irvine). New York: Wiley.

Kendall, R.J. and Akerman, J. (1992). Terrestrial wildlife exposed to agrochemicals: an ecological risk assessment perspective. *Environ. Toxicol. Chem.* 11 (12): 1727–1749. https://doi.org/10.1002/etc.5620111206.

LaLone, C.A., Villeneuve, D.L., Lyons, D. et al. (2016). Editor's highlight: sequence alignment to predict across species susceptibility (SeqAPASS): a web-based tool for addressing the challenges of cross-species extrapolation of chemical toxicity. *Toxicol. Sci. Off. J. Soc. Toxicol.* 153 (2): 228–245. https://doi.org/10.1093/toxsci/kfw119.

LaLone, C.A., Villeneuve, D.L., Doering, J.A. et al. (2018). Evidence for cross species extrapolation of mammalian-based high-throughput screening assay results. *Environ. Sci. Technol.* 52 (23): 13960–13971. https://doi.org/10.1021/acs.est.8b04587.

LaLone, C.A., Basu, N., Browne, P. et al. (2021). International consortium to advance cross-species extrapolation of the effects of chemicals in regulatory toxicology. *Environ. Toxicol. Chem.* 40 (12): 3226–3233. https://doi.org/10.1002/etc.5214.

LaLone, C.A., Blatz, D.J., Jensen, M.A. et al. (2023). From protein sequence to structure: the next frontier in cross-species extrapolation for chemical safety evaluations. *Environ. Toxicol. Chem.* 42 (2): 463–474. https://doi.org/10.1002/etc.5537.

Landis, W.G. (2026). Chapter 7: Management of environmental systems using cumulative risk assessment. In: *Solving Complex Environmental Challenges from Multiple Stressors* (ed. L.A. Kapustka, W.G. Landis, D.J. Ostrach, and C.A. Irvine). New York: Wiley.

Landis, W.G., Markiewicz, A.J., Ayre, K.K. et al. (2017). A general risk-based adaptive management scheme incorporating the Bayesian network relative risk model with the South River, Virginia, as case study. *Integr. Environ. Assess. Manag.* 13 (1): 115–126. https://doi.org/10.1002/ieam.1800.

Langan, L.M., Paparella, M., Burden, N. et al. (2023). Big question to developing solutions: a decade of progress in the development of aquatic new approach methodologies from 2012 to 2022. *Environ. Toxicol. Chem.* https://doi.org/10.1002/etc.5578.

Lewis, N.S., Marois, D.E., Littles, C.J., and Fulford, R.S. (2020). Projecting changes to coastal and estuarine ecosystem goods and services—models and tools. In: *Ecosystem-Based Management, Ecosystem Services and Aquatic Biodiversity: Theory, Tools and Applications* (ed. T.G. O'Higgins, M. Lago, and T.H. DeWitt), 235–253. Cham: Springer International Publishing. https://doi.org/10.1007/978-3-030-45843-0_12.

Lillicrap, A., Belanger, S., Burden, N. et al. (2016). Alternative approaches to vertebrate ecotoxicity tests in the 21st century: a review of developments over the last 2 decades and current status. *Environ. Toxicol. Chem.* 35 (11): 2637–2646. https://doi.org/10.1002/etc.3603.

Margiotta-Casaluci, L., Luigi, O.S.F., and Winter, M.J. (2023). Cross-species extrapolation of biological data to guide the environmental safety assessment of pharmaceuticals—the state of the art and future priorities. *Environ. Toxicol. Chem.* https://doi.org/10.1002/etc.5634.

Mayasich, S.A., Goldsmith, M.R., Mattingly, K.Z. et al. (2023). Combining in vitro and in silico new approach methods to investigate type 3 iodothyronine deiodinase chemical inhibition across species. *Environ. Toxicol. Chem.* 42 (5): 1032–1048. https://doi.org/10.1002/etc.5591.

NRC (2007). *Toxicity Testing in the 21st Century: A Vision and a Strategy*. Washington, DC: National Academies Press. https://doi.org/10.17226/11970.

ORISE (2023). Oak Ridge institute for science and education (ORISE). https://orise.orau.gov/index.html (accessed 27 March 2026).

Raimondo, S., Mineau, P., and Barron, N.G. (2007). Estimation of chemical toxicity to wildlife species using interspecies correlation models. *Environ. Sci. Technol.* 41 (16): 5888–5894. https://doi.org/10.1021/es070359o.

Rattner, B.A., Bean, T.G., Beasley, V.R. et al. (2023). Wildlife ecological risk assessment in the 21st century: promising technologies to assess toxicological effects. *Integr. Environ. Assess. Manag.* https://doi.org/10.1002/ieam.4806.

Richard, A.M., Huang, W.R.S., Shinn, P. et al. (2021). The Tox21 10K compound library: collaborative chemistry advancing toxicology. *Chem. Res. Toxicol.* 34 (2): 189–216. https://doi.org/10.1021/acs.chemrestox.0c00264.

Sample, B.E., Johnson, M.S., Hull, R.N. et al. (2023). Key challenges and developments in wildlife ecological risk assessment: problem formulation. *Integr. Environ. Assess. Manag.* https://doi.org/10.1002/ieam.4710.

Schaupp, C.M., LaLone, C.A., Blackwell, B.R. et al. (2023). Leveraging ToxCast data and protein sequence conservation to complement aquatic life criteria derivation. *Integr. Environ. Assess. Manag.* 19 (1): 224–238. https://doi.org/10.1002/ieam.4617.

Socol, Y., Shaki, Y.Y., and Yanovskiy, M. (2019). Interests, bias, and consensus in science and regulation. *Dose-Response* 17 (2): 1559325819853669. https://doi.org/10.1177/1559325819853669.

Stahl, R.G. Jr., (2026). Chapter 9: Reflections from an environmental toxicologist working in a large company. In: *Solving Complex Environmental Challenges from Multiple Stressors* (ed. L.A. Kapustka, W.G. Landis, D.J. Ostrach, and C.A. Irvine). New York: Wiley.

Stelzenmüller, V., Coll, M., Cormier, R. et al. (2020). Operationalizing risk-based cumulative effect assessments in the marine environment. *Sci. Total Environ.* 724 (July): 138118. https://doi.org/10.1016/j.scitotenv.2020.138118.

Sun, X., Li, B., Han, F. et al. (2019). Impacts of arsenic and antimony co-contamination on sedimentary microbial communities in rivers with different pollution gradients. *Microb. Ecol.* 78 (3): 589–602. https://doi.org/10.1007/s00248-019-01327-5.

Suter, G. (2023). Successful innovation in environmental assessment. *Integr. Environ. Assess. Manag.* 19 (2): 556–557. https://doi.org/10.1002/ieam.4738.

Thomas, R.S., Bahadori, T., Buckley, T.J. et al. (2019). The next generation blueprint of computational toxicology at the U.S. environmental protection agency. *Toxicol. Sci. Off. J. Soc. Toxicol.* 169 (2): 317–332. https://doi.org/10.1093/toxsci/kfz058.

USEPA (2021). *A Proof-of-Concept Case Study Integrating Publicly Available Information to Screen Candidates for Chemical Prioritization under TSCA*. Washington, DC. EPA/600/R-21/106. https://cfpub.epa.gov/si/si_public_pra_view.cfm?dirEntryID=349776&Lab=CCTE: U.S. Environmental Protection Agency (accessed 27 March 2026).

USEPA (2022). Sustainable and healthy communities (SHC) – strategic research action plan, fiscal years 2023–2026. EPA/600/R-22/241 | October 2022. https://www.epa.gov/system/files/documents/2022-10/SHC%20FY23-26%20StRAP_EPA-ORD_October%202022_508.pdf (accessed 26 March 2026).

USEPA Engineering Forum Issue Paper (2017). Ecosystem services at contaminated site cleanups. EPA 542-R-17-004. https://www.epa.gov/sites/default/files/2017-09/documents/ecosystem_services_at_contaminated_site_cleanups_ef_issue_paper.pdf (accessed 27 March 2026).

USEPA, OCSPP (2018). Strategic plan to promote the development and implementation of alternative test methods within the TSCA program. https://www.epa.gov/sites/default/files/2018-06/documents/epa_alt_strat_plan_6-20-18_clean_final.pdf (accessed 26 March 2026).

USEPA, OLEM (2008). ECO update/groundwater forum issue paper: evaluating ground-water/surface-water transition zones in ecological risk assessments.

Reports and assessments. https://www.epa.gov/remedytech/eco-update-groundwater-foum-issue-paper-evaluating-ground-watersurface-water-transition (accessed 27 March 2026).

USEPA, ORD (2018). PVIScreen. Data and tools. 20 February 2018. https://www.epa.gov/land-research/pviscreen (accessed 26 March 2026).

USEPA, ORD (2022). Strategic research action plans fiscal years 2023–2026. https://www.epa.gov/research/strategic-research-action-plans-fiscal-years-2023-2026 (accessed 26 March 2026).

Welch, S.A. (2026). Chapter 5: Addressing the challenges of pharmaceutical pollution. In: *Solving Complex Environmental Challenges from Multiple Stressors* (ed. L.A. Kapustka, W.G. Landis, D.J. Ostrach, and C.A. Irvine). New York: Wiley.

Williams, K.C. and Hoffman, J.C. (2020). Remediation to restoration to revitalization: engaging communities to support ecosystem-based management and improve human wellbeing at clean-up sites. In: *Ecosystem-Based Management, Ecosystem Services and Aquatic Biodiversity* (ed. T.G. O'Higgins, M. Lago, and T.H. DeWitt), 543–559. Cham: Springer International Publishing. https://doi.org/10.1007/978-3-030-45843-0_27.

Williams, A.J., Grulke, C.M., Edwards, J. et al. (2017). The CompTox chemistry dashboard: a community data resource for environmental chemistry. *J. Chem.* 9 (1): 61. https://doi.org/10.1186/s13321-017-0247-6.

APPENDIX

Results of Author Survey

Questions to Authors:

1) Are you including a discussion of challenges to adopting the tools/approaches mentioned in your chapter?
   If so, can you share a brief synopsis of those challenges with us?
   If not, can you please share your opinions on the main (e.g., top 3–5) challenges to getting new tools and approaches actually used by decision makers? And any opinions on overcoming each challenge?
2) Beyond your specific chapter, do you have examples that you could share of new/emerging tools that have:
   Been mainstreamed successfully to be fully considered as a basis for decisions? And why were they successful? or
   Hit a barrier to acceptance and use?

Answers:

Mary A. Ottinger:
28 July 2023

1) I would like to reinforce Ralph's response.
   Getting and using the tools, especially making them field friendly, is a big challenge. This includes convincing the field biologists to use the tools in a way that makes the interpretation of findings both field and public dissemination friendly. In addition, the multiple stressor condition that all living organisms encounter is no small matter; capturing this aspect is challenging.
2) Again, I agree with Ralph. It's critical to deal with the complexities of stressors that are overlaid upon each other and consider factors that exacerbate the potential for adverse effects. Structured approaches are useful to organize a stepwise process to address complex challenges and reassess approaches to adjust as needed, such as with adaptive management.

Ralph Stahl:
28 July 2023

1) Yes. I'm including a discussion of challenges to adopting the tools/approaches mentioned in my chapter.
   I think the main challenge is having enough time and evidence to show that the new tool or approach is not a way to "hide the ball." I know that Wayne might have addressed this in some of his writings as he was involved in one of our cases where we had to overcome skepticism on using a multiple-stressor approach which might suggest the chemical(s) in question were not the greatest source of ecological risk.
2) While the six manuscripts have just been submitted to IEAM, I believe they will show clearly that some elements of climate change model projections can be incorporated into ecological risk assessments and at spatial scales appropriate for making risk management/climate adaptation decisions. These manuscripts resulted from a SETAC Pellston workshop held in June 2022, and I think (and hope) they will dislodge ecological risk assessors and decision makers from their singular focus on chemicals … in the absence of what stress climate change could add to ecological risks.

Wayne Landis:
28 July 2023

1) Our chapter deals with mixtures and uses Bayesian networks (BNs) as a framework. Every risk assessment paper my lab has published since 2013 has used BNs for calculations. In most cases, we have been invited to work with the

agency or industry. In the case that Ralph talks about regarding the South River, we did contribute to the decision-making process, but it took a lot of explanation and examples to demonstrate how the process works. We published those papers in 2017 in IEAM and Risk Analysis. We are about to start submitting the work on the California Delta risk assessment – a huge project with decades worth of data. As you probably know, there are labs using BNs with the EPA (see Carriger publications) in Europe (J. Moe and others), South Africa (Gordon O'Brien), and others. Some of these papers have been selected as best papers in IEAM or as runner ups. As Ralph mentioned, there are six papers submitted to IEAM using BNs to describe the risks due to climate change. So it has been successful in that manner.

However, there is still huge pushback. Suter just published a learned discourse in IEAM regarding the adoption of new methods (see attached). The EU still uses the species sensitivity distribution for decision making with the calculation of a hazard quotient as the final result. There have been a number of so-called risk assessments for microplastics that use an SSD approach using NOECs as the endpoint to estimate a water quality target for these materials. There is not even a consideration to use a cause–effect conceptual model with exposure–response modeling and specific endpoints being applied.

Although conceptually problematic, the use of the SSD has become a regulation and required. In pesticide regulation, the calculation of an HQ still seems to be the endpoint even if a probabilistic tool has been used to get there. Nothing like condensing all the information into an index or even simple distribution, where the critical information on the different types of uncertainties are lost.

2) BN approaches are the norm in the US Forest Service for some time now. But it is a very different agency. Bruce Marcot pioneered the approach for wildlife management years ago, and those publications are key. Fisheries management in Finland has been done using BNs for some years, but again that is because of a few key players. In both of these cases, there was a small group of people that advanced the ideas along with a small group that adopted them.

   Here are the challenges.

   The field of environmental toxicology is slow to adopt modern data analysis and other tools. Most papers in the field still use an AVOVA multiple comparisons experimental design, although the severe limitations of the approach have been extensively documented. As someone who connects with people in the synthetic biology and molecular worlds, I am amazed how slowly our field adopts change.

   The training of new people to the field has lacked in providing a solid foundation in data analysis and experimental design. In other words, academia has been teaching the past. We cannot seem to be able to adopt even simple tools such as regression to describe exposure–response, or even the drc package in

R (Analysis of Dose-Response Curves) and many other commercial packages that are available. My department at WWU has a data analysis series with a certificate that many seniors and almost all MS students take. Yet papers are published in the SETAC journals using archaic methods all the time.

12

Going Forward

Lawrence A. Kapustka

LK Consultancy, Diamond Valley, AB, Canada

Introduction

Advances in computational capacity, storage, searchability, and pattern recognition in recent years have opened possibilities to gain insight into complex data systems. Especially since the passage of the National Environmental Protection Act in 1971, massive datasets have been developed on a variety of environmental entities. The types of data include concentrations of various chemicals in air, water, soil, sediment, and biota. During this same period, resource agencies have tracked populations of fish and wildlife, surveyed aquatic and marine systems, and monitored forest and other crop production. Health officials have documented disease parameters in humans, domestic livestock, and wildlife, much of which is now discussed under the One Health paradigm. Additionally, the basic sciences of ecology, landscape ecology, systems ecology, theoretical ecology, political science, and sociology have contributed new insights as to how each of these fields offers great potential for advancing our collective knowledge of social-ecological systems. But we also need to be aware of the quality and relevance of the output generated through the exploration of big data and be cautious.

The first and perhaps the largest challenge with found data, that is data developed for different usage, is qualifying the reliability for the new usage. This lesson was brought home to many during the US Environmental Protection Agency (USEPA) project to establish soil screening values (ecological soil screening levels or Eco-SSLs) to use in the Superfund Program (Ecological Soil Screening Level [Eco-SSL] Guidance and Documents | USEPA accessed 28 November 2023). One objective of the Eco-SSL effort was to establish a common suite of published studies on selected analytes from roughly the period of 1988 through 2000 so that practitioners would not need to repeat extensive literature searches and reviews.

Another was to set an agreed concentration for each target analyte such that if the concentration at a site was below the threshold, the analyte could be dropped from further consideration at that site. The pattern that emerged for polyaromatic hydrocarbons (PAHs) was similar for other target analytes. What follows are the key observations.

Standard operating procedures were developed to govern electronic literature searches, codify the criteria used to judge acceptance of studies, provide guidelines for evaluating data from accepted studies, and define steps for calculating the values to be used as screening concentrations. After assembling lists of potentially useful papers, a systematic review was done to ensure that only studies with primary data (i.e., elimination of secondhand summaries, only modelling results, etc.) would be analyzed. Each study within a paper was then scored against a set of acceptance criteria, which included the following:

- The paper had to be a primary source of literature.
- Adverse effects due to a single chemical stressor had to be reported. The toxicant chemical form and concentrations were reported.
- The test medium used in the study was either natural or artificial soil. The organic matter content was reported and was <10%.
- Soil pH was reported and was within the range of 4.0–8.5.
- Control treatments were reported.
- Duration of the study was either reported or inferred by virtue of referencing a standard test method, which has a specified exposure period.
- For laboratory studies, at least three treatment concentrations including the controls were used.
- Biological effects were reported for ecologically relevant endpoints (e.g., growth, reproductive endpoints).

Nearly 100 papers on PAHs effects on plants or invertebrates were examined in detail (see Kapustka 2004). Only 3.2% of the studies reported in those papers had information that was useful in setting Eco-SSLs as judged against the aforementioned criteria. In the end, only 3 of the 16 target analyte PAHs (anthracene, naphthalene, and acenaphthene) had reports of phytotoxicity studies using terrestrial species. None was sufficient for deriving plant-SSLs. There was insufficient information to set Eco-SSLs for invertebrates or wildlife.

The conclusion drawn from this, and other comprehensive literature reviews, is not that the papers were necessarily flawed, though some were, but rather that data were collected to answer different questions. In some cases, papers were missing critical information that the authors undoubtedly had but were limited in what they could put into the papers to conform with publisher guidelines. The ability now to include supplemental data with published papers alleviates some of this problem, but the central issue remains that data collected for one purpose often is inadequate for a different analysis.

Even after data entries are gleaned to eliminate duplicates, entry errors, and the like, additional questions should be addressed. These include (in no particular order of importance) the following:

- Are the data from one study or one site comparable to those of another study or site?
- Were similar detection limits used across all studies?
- Were data for one parameter co-located (time and space) with other data? For example, for a site characterization, were water quality data and invertebrate or fish sampling done at the same time and place? For vegetation studies, were soil samples and plant samples co-located?
- In any study, was the sampling plan capable of producing data that were representative of the site? Have sufficient statistical power to be meaningful? Have acceptable precision and accuracy ascribed to the study plan?
- Related to these, were data collected from a recognizably common statistical population, or did the samples come from two or more statistical populations? For example, vegetation data coming from areas with distinctly different soil types would most likely require postclassification into two statistical groups rather than be treated as one group.

Rapidly emerging artificial intelligence (AI) technologies have the capacity to sort through these nuances in data and find new insights through pattern recognition. However, at least for the near future, it will require careful interrogation of patterns that are revealed. Are the results showing interesting correlations or might there be underlying causal factors that explain the recognized patterns.

Finally, as with all visualizations, care needs to be taken in considering the resolution of the patterns. How accurate are the boundaries that are depicted? What is the fidelity of the mapping units? Do all the challenges of spatial statistics that were prominent in the 1980s and 1990s apply to depictions of the patterns derived from big data?

Although we have witnessed incredible advances in the depiction of social-ecological systems over our decades in the field, there remain many troubling challenges. As noted earlier (Chapter 2), there has been little progress in dealing with contrived limitations in the realm of ecological risk assessment. Even in this collection of chapters, we see that some disciplines appear to be stuck with reliance on flawed methods (i.e., no observable effects level or concentration (NOEL, NOEC)s and lowest observable effects level or concentration (LOEL or LOEC) s and comparable point estimates) because "those are the only data available." Similarly, the foundational work of Kay, Hollings, Allen, Funtowicz, Ravetz and their colleagues describing the dynamics that operate in systems ecology have largely been ignored. A central premise of postnormal science (Funtowicz and Ravitz 1993) is that systems complexity, characterized by multiple nested feedback loops, precludes assignment of causality. Yet, our institutions, especially those dealing with environmental law, tilt heavily toward a requirement of an established causal linkage for enforcement. Almost all monitoring programs are designed to focus on rapid response endpoints that are believed to reflect a direct link to an action (release, or practice). This, while systems ecology has for several decades informed us that slow acting feedback loops are more determinative (e.g., climate change) than the faster acting feedback loops that are typically the focus of monitoring programs.

Collectively, there are reasons for optimism that we can gain insights from past efforts in basic and applied sciences. Impressive advances have been realized, fueled by technologies. For many of the contemporary challenges, such as declining biodiversity, fragmentation of critical habitats, climate change, human population growth, and pandemics, there is little time to waste. The question is whether there is societal will and the professional responsivity to move beyond our comfort zones.

And this is now all in flux as the policies and practices of the Trump administration have taken direct aim at the institutions that rely on science. The gutting of research at various institutions (e.g., National Institute of Health (NIH), the Environmental Protection Agency (EPA), and the Department of the Interior (DOI)) coupled with misinformation and disinformation campaigns do not bode well for the future. Treatises by Timothy Snyder (*On Tyranny* 2017 and *On Freedom* 2024) along with the in-depth work on authoritarianism by Ruth Ben-Ghiat (*Strongmen: Mussolini to the Present* 2020) warn of massive societal change that would not be favorable to advances in environmental sciences.

While there is political turmoil in so many regions, the rapid rise of AI is likewise becoming disruptive. Despite the great potential to achieve substantive breakthroughs in pattern recognition that could markedly improve forecasting of weather events and other ecological influences, Yuval Noah Harari in *Nexus* (2024) sounded the alarm regarding AI-generated algorithms that could soon be beyond the capacity of humans to decipher how they function. In short, we would lose the capacity to know if what is being delivered is verifiable or falsifiable. In other words, the output could soon be beyond the realm of science as we know it. If we are to succeed in assessing and managing contemporary challenges in our social-ecological setting, great courage and vigilance will be required.

References

Ben-Ghiat, R. (2020). *Strongmen: Mussolini to the Present*, 384. New York: WW Norton.

Funtowicz, S.O. and Ravitz, J.R. (1993). Science for the post-normal age. *Futures* 245: 739–955.

Harari, Y.N. (2024). *Nexus: A Brief History of Information Networks from the Stone Age to AI*, 492. New York: Random House.

Kapustka, L.A. (2004). Establishing eco-SSLs for PAHs: lessons revealed from a review of literature on exposure and effects to terrestrial receptors. *Hum. Ecol. Risk. Assess.* 10: 185–205.

Snyder, T. (2024). *On Freedom*, 345. New York: Crown.

Snyder, T. (2017). *On Tyranny*, 126. New York: Crown.

Index

a

17-alpha ethinylestradiol 78
acetylcholinesterase (AChE) 17, 74–77, 99, 101, 102
adverse outcome pathways (AOPs) 24–26, 33, 192, 205
alkylphenols 78
American Society for Testing and Materials (ASTM) 11, 14, 17–19, 21, 28, 37, 40, 42, 43, 141, 182, 183, 192
Ankistrodesmus 25, 167, 169
apoptosis 74, 77
arsenic 72, 216
atrazine 19

b

Baird, D.J. 2
Bayesian 6, 17, 18, 21–23, 26, 28, 34, 35, 48, 54, 55, 57, 59, 61, 62, 64–66, 91, 96, 102, 111–114, 159, 164, 171, 172, 174, 175, 178, 181–183, 196, 202, 214, 218
bifenthrin 22, 23, 34, 100–103
big data 8, 25, 30, 221, 223
bioaccumulation 29, 30, 37, 73, 82, 125, 129, 132, 141, 142, 144, 189, 206, 213
bioassays 7, 8, 13, 17, 18, 20, 30, 33, 36
biodiversity 11, 47, 48, 62, 176, 224
biomarkers 25, 33, 34, 73, 74, 76, 79, 83, 87, 132, 143, 207
biotic ligand model (BLM) 22
BN relative risk model (BN-RRM) 28, 92, 93, 96, 171, 176–178, 202
Burton, G.A. Jr. 2

c

cadmium 21, 22, 42, 75, 77
carbamates 75, 76
Carson, R. xxix
Centers for Disease Control and Prevention 26
Central Valley Regional Water Quality Control Board 29, 72, 84
Chemical Abstract Services (CAS) 35
Chesapeake Bay 82, 88
Chinook 7, 34, 94, 96–99, 106, 109, 110, 113, 171, 183
Chinook salmon 94, 97, 110, 113, 183
Chlorella 167
Chlorpyrifos 22, 34, 42, 88, 101–103, 107, 108, 113, 182
Christina River 152, 153
chromium 21
Clean Air Act 5, 7

climate change 2, 10, 13, 14, 21, 30, 47–49, 52, 54, 62, 64–67, 93, 109–111, 113, 125, 138, 147, 155, 157, 158, 160, 161, 163, 170, 180, 181, 183, 202, 203, 218, 219, 224
community conditioning hypothesis 38, 44, 166, 169, 170, 182
Comprehensive Environmental Response, Compensation, and Liability Act (CERCLA) 17, 18, 32, 160
Conditional Probability Tables (CPTs) 7, 54, 104
Council on Environmental Quality 5
cypermethrin 23
cytochrome P4501A (CYP1A) 17, 74, 75, 78, 85, 87, 89

d

2,4-D 19
Daphnia 25, 31, 166, 167
Deep-Water Horizon Oil Spill 26, 27
Delaware 8, 88, 151–153, 157, 160
deltamethrin 23
diazinon 34, 101–103, 107, 108
dichlorodiphenyltrichloroethane (DDT) 15, 17, 72, 73, 79, 115
diclofenac 48, 58, 67, 142
dieldrin 72
dioxins 19, 72, 74, 75
DuPont 8, 16, 147, 150–158

e

ecosystem services 11, 48, 62, 63, 94, 181, 213, 214
EcoToxChip 24, 28, 29, 31, 138
Endangered Species Act 7, 134, 206
equilibrium partitioning 18, 22, 29
ESFA 126
7-ethoxyresorufin-O-deethylase (EROD) 6, 18, 74, 75, 78, 79, 84, 86, 87, 89
esfenvalerate 23
European flounder 78
European Free Trade Association 56
European Union 18, 56, 64, 118, 125, 126
exposome 5, 16, 21, 25–27, 31

f

fate and transport 2, 20, 117
Federal Environmental Pesticide Control Act 7
Federal Insecticide, Fungicide, and Rodenticide Act 18, 19
Federal Land Policy and Management Act 7
Ferenc, S.A. 2
Foran, J.A. 2
Forman, R.T.T. xxxi
Funtowicz, S.O. xxxii

g

gas chromatography 14
genomics 24, 144
Godron, M. xxxi
grey mullet 78, 85
Gulf of Mexico 26

h

habitat suitability 129
heart 39, 75
hepatopancreatic vessels 75
hierarchical patch dynamics 163, 184
high-performance liquid chromatography 14, 18
Hurlbert, S.H. 11, 20, 30
Hyalella azteca 16

i

immunohistochemical (IHC) 75
in situ toxicity 20

j

Jet A 166–168
JP-4 166, 167, 182
JP-8 25, 166–169, 182

k

Kay, J. 32
ketoconazole antifungals 75

l

lambda-cyhalothrin 23
lamellar pillar cells 75
Landscape Ecology 32, 141
largemouth bass 72
lead 11, 21, 22, 25, 29, 33, 35, 41, 45, 74, 79, 130, 137–139, 142, 200, 205
longfin smelt 98
Lyngbya 167

m

malathion 19, 34, 101–103
mass spectrometry 14, 18
maternal transfer 73, 76, 77, 79, 80, 82, 126
mercury 13–15, 22, 34, 72, 75, 77, 84, 88, 93, 100–102, 111, 112, 138, 153–155, 159, 160, 182
metabolomics 24
metallothionein 18, 36, 74, 87–89
methanol 75
2-methlypropane 166
microcosm 25, 34, 165–169, 182, 183
mixed flask culture (MFC) 25, 34, 165
Mokelumne River 79, 80, 95
mummichogs 79, 87
myclobutanil 34, 101
mycorrhizal fungi 38

n

National Environmental Protection Act 5, 6, 9, 18, 30, 31, 221
National Oceanic and Atmospheric Administration (NOAA) 18, 152, 153
Natural Resources Defense Council 154, 155
Nelson, G., xxix
Newport 8, 151–154, 157, 160
nickel 21, 22

o

Occupational Safety and Health Act 7
One Health 5, 16, 21, 25–27, 31, 221
Organisation for Co-operation and Development (OECD) 18, 128, 132, 134–136, 143, 145
organophosphate insecticides 16
Ostracod 167
otolith 72, 80, 88
oxidative stress 74, 77, 89

p

pelagic organism decline (POD) 13, 71
Pellston Workshop 2, 4
people skills 147, 154, 199, 210
per- and polyfluoroalkyl substances (PFAS) 15, 16, 19, 29, 206
perfluorooctanesulfonic acid (PFOS) 15, 19, 33
perfluorooctanoic acid (PFOA) 15, 33
permethrin 23
pharmaceutical 21, 22, 47–51, 54–57, 59–65, 67, 68, 196, 204, 217
phthalates 78
pollution-induced community tolerance (PICT) 38
polychlorinated biphenyls (PCBs) 15, 19, 72–75, 78, 79, 96, 115, 141
polychlorinated dibenzofurans 74
polycyclic aromatic hydrocarbons (PAHs) 15, 19, 72, 74, 75, 78, 84, 87, 92, 170, 171, 222, 225

Port Valdez 8, 114, 165, 170, 172–174, 184
Predicted or Measured Environmental Concentration (PEC or MEC) 53
Predicted No Effect Concentration (PNEC) 53
proteomics 24
Pseudokirchneriella subcapitata 19
pseudoreplication 20
pyrethrins 16
pyrethroids 16, 23, 78

q

quantitative polymerase chain reaction (qPCR) 24

r

rainbow trout 31, 78, 83, 85, 87, 89
reactive oxygen species (ROS) 19, 74, 84
Record of Decision (ROD) 19, 152, 196
Red Book 163
relative risk model (RRM) 22, 26, 28, 65, 66, 91, 92, 96, 112, 113, 159, 170–172, 176, 182, 183, 214
remediation 8, 12, 13, 147–153, 155–157, 180, 187, 188, 200, 208
resiliency 10, 32
Resource Conservation and Recovery Act (RCRA) 7, 8, 19, 148, 149, 151, 153–156
Resource Recovery Act 7
restoration 13, 39, 97, 147, 150–154, 157, 159, 160, 179, 180, 184, 187, 188, 217
Rittel, H. xxxii

s

Sacramento River 24, 79, 80, 83, 84, 94, 95, 103
Sacramento-San Joaquin River Delta 33, 71, 83
Safe Drinking Water Act 7
San Francisco Bay Estuary xxxii
San Joaquin River xxxii, 71, 79, 80, 83–85, 94
Securities and Exchange Commission 20, 148, 161
Selanastrum 167
Selenastrum capricornutum 19
selenium 13, 14, 33, 36
Silent Spring xxix, 5, 9
silver 21, 22
social-ecological systems 4, 7, 9–11, 30, 33, 40, 187, 221, 223
South River 65, 93, 96, 112, 113, 153–155, 159, 171, 179, 180, 215, 219
sRNA 8
standard aquatic microcosm (SAM) 25, 34, 165, 166, 168
steelhead 98
steroids 78
Stockholm Resilience Centre 19, 47
striped bass13 71–77, 79–84, 87, 88, 98, 100
sunshine bass 79, 87
Superfund 8, 18, 19, 31, 148–152, 156, 160, 195, 196, 221
Surface Mining Control and Reclamation Act 7
synchrotron 14, 28, 29

t

threadfin shad 98
toxaphene 72
ToxCast 24, 216
Toxic Substances Control Act 7, 188
toxicity reference values 131, 137, 138, 140, 143
transcriptomics 24, 30
triadimefon 34, 101
tributyltin 78, 87

Trichodina 81
Trichodiniasis 81
triclosan 78, 86
TritonX-100 75
TRV 19, 129, 137
Tween 80 75

u

UN Department of Economic and Social Affairs 50, 68
UNESCO 30
United Kingdom 8, 151, 155, 156
Upper San Francisco Estuary 7, 22, 34, 91, 94–97, 101, 103, 105, 107, 113
Urban Wastewater Treatment Directive 56
United States Environmental Protection Agency (USEPA) 5, 8, 11, 12, 14–16, 18–24, 29, 31–33, 36, 45, 46, 98, 100, 114, 118, 120, 122, 128, 135, 136, 145, 148, 152, 154–156, 160, 183, 188, 189, 193, 195, 196, 198, 204, 208, 214, 216, 217, 221

v

Virginia 8, 20, 65, 93, 112, 113, 153–155, 171, 182, 215
vitellogenin 20, 74, 78, 86, 89

w

Water Pollution and Control Act Amendments 7
Waynesboro 8, 153–155, 171
Webber, M. xxxii
weight of evidence 20
white catfish 72
WHO 20, 47, 51, 52, 69
wicked problems 33, 47, 186

x

X-ray fluorescence 15

z

zebrafish 24, 31, 84–86, 89
zinc 21, 22, 29, 31, 77

Wiley The manufacturer's authorized representative according to the EU General Product Safety Regulation is Wiley-VCH GmbH, Boschstr. 12, 69469 Weinheim, Germany, e-mail: Product_Safety@wiley.com.

Printed and bound by CPI Group (UK) Ltd, Croydon, CR0 4YY
02/09/2026
02207528-0002